Catenanes, Rotaxanes, and Knots

ORGANIC CHEMISTRY

A SERIES OF MONOGRAPHS

ALFRED T. BLOMQUIST, *Editor* ·

Department of Chemistry, Cornell University, Ithaca, New York

1. Wolfgang Kirmse. CARBENE CHEMISTRY, 1964; 2nd Edition, *In preparation*

2. Brandes H. Smith. BRIDGED AROMATIC COMPOUNDS, 1964

3. Michael Hanack. CONFORMATION THEORY, 1965

4. Donald J. Cram. FUNDAMENTAL OF CARBANION CHEMISTRY, 1965

5. Kenneth B. Wiberg (Editor). OXIDATION IN ORGANIC CHEMISTRY, PART A, 1965; PART B, *In preparation*

6. R. F. Hudson. STRUCTURE AND MECHANISM IN ORGANO-PHOSPHORUS CHEMISTRY, 1965

7. A. William Johnson. YLID CHEMISTRY, 1966

8. Jan Hamer (Editor). 1,4-CYCLOADDITION REACTIONS, 1967

9. Henri Ulrich. CYCLOADDITION REACTIONS OF HETEROCUMULENES, 1967

10. M. P. Cava and M. J. Mitchell. CYCLOBUTADIENE AND RELATED COMPOUNDS, 1967

11. Reinhard W. Hoffman. DEHYDROBENZENE AND CYCLOALKYNES, 1967

12. Stanley R. Sandler and Wolf Karo. ORGANIC FUNCTIONAL GROUP PREPARATIONS, VOLUME I, 1968; VOLUME II, *In preparation*

13. Robert J. Cotter and Markus Matzner. RING-FORMING POLYMERIZATIONS, PART A, 1969; PART B, *In preparation*

14. R. H. DeWolfe. CARBOXYLIC ORTHO ACID DERIVATIVES, 1970

15. R. Foster. ORGANIC CHARGE-TRANSFER COMPLEXES, 1969

16. James P. Snyder (Editor). NONBENZENOID AROMATICS, I, 1969; II, *In preparation*

17. C. H. Rochester. ACIDITY FUNCTIONS, 1970

18. Richard J. Sundberg. THE CHEMISTRY OF INDOLES, 1970

19. A. R. Katritzky and J. M. Lagowski. CHEMISTRY OF THE HETEROCYCLIC N-OXIDES, 1970

20. Ivar Ugi (Editor). ISONITRILE CHEMISTRY, 1971

21. G. Chiurdoglu (Editor). CONFORMATIONAL ANALYSIS, 1971

22. Gottfried Schill. CATENANES, ROTAXANES, AND KNOTS, 1971

Catenanes, Rotaxanes, and Knots

GOTTFRIED SCHILL

Chemisches Laboratorium
University of Freiburg
Freiburg, Germany

Translated by
J. BOECKMANN

Chemisches Laboratorium
University of Freiburg
Freiburg, Germany

ACADEMIC PRESS   **1971**   **New York and London**

ACADEMIC PRESS, INC.
111 Fifth Avenue, New York, New York 10003

United Kingdom Edition published by
ACADEMIC PRESS, INC. (LONDON) LTD.
Berkeley Square House, London W1X 6BA

LIBRARY OF CONGRESS CATALOG CARD NUMBER: 78-127702

PRINTED IN THE UNITED STATES OF AMERICA

Contents

v

Preface

Catenanes and rotaxanes differ from all other organic compounds synthesized thus far in that molecular subunits are linked mechanically. For this reason, these structures have for a long time attracted considerable interest.

Recent investigations have led to a series of syntheses for this class of compounds. The synthetic work on catenanes and rotaxanes has opened the door for even more complex structures, such as, for example, the knot. This latter type of compound is included in this monograph because the synthetic methods, as well as the theoretical problems, are similar to those of catenanes and rotaxanes.

I agreed to write this monograph although I am aware that on the basis of the work carried out thus far it can only be a collection and arrangement of synthetic information. The reader will search in vain for information concerning physical and chemical properties of this class of compounds, since, with the exception of a few investigations such as the studies on mass spectrometric fragmentation patterns, most of this type of research remains to be done. Although from the experimental point of view the research in this area is often arduous, it will be interesting to investigate the manner of interaction between the molecular subunits in catenanes in the solid, liquid, and gaseous states, as well as the stereochemistry and the other physical and chemical properties related to the mechanical linkages in these compounds.

A large part of the synthetic work dealt with in this monograph is contained in unpublished dissertations or, as in the case of my research group, in dissertations not yet finished. For this reason, I had to evaluate many primary sources and to reproduce many experimental details.

I tried to arrange the topics according to the theories that finally led to the synthesis of catenanes and rotaxanes. This organization often led to digressions from the original concept; digressions which, although essential to the monograph, frequently reveal new synthetic possibilities. The clarity of the text may have suffered thereby, but I hope that those who work in this area in the future may thereby find it easier to acquaint themselves with the difficulties encountered in the experimental work.

Throughout the writing of the manuscript, numerous colleagues were of assistance to me and I wish to express my gratitude to all of them. Special thanks are due to Mr. Juan Boeckmann, MSc., who carried out the English translation and organized numerous parts of the manuscript. I also thank Dr. G. Isele who proofread the entire manuscript.

Catenanes, Rotaxanes, and Knots

1

History

The synthesis of structures as yet not found in nature has always been a certain challenge to chemists. Therefore, the idea of synthesizing molecules composed of separate entities which are mechanically linked to one another has aroused, even if only casually, the interest of many investigators. The above idea is embodied in the structure of a linked chain or catenane[1] **1** (Latin: *catena* = the chain) in which macrocyclic molecules are held together only mechanically without the aid of a chemical bond. Since for a long time these ideas could not be associated with tangible facts, they did not enter the chemical literature.

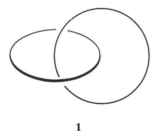

1

According to Prelog this type of compound was mentioned by Willstätter in a seminar at Zürich in the years between 1900 and 1912, that is, long before macrocycles were even known.[2, 3] These ideas could not have been realistic, however, since little was known about the size and spatial arrangement of the rings required.

With the discovery of cyclic polymers in synthetic polycondensates,[4] speculations about the existence of catenanes gained new impetus. Following the law of statistics it could be expected that, besides the cyclic monomers and oligomers, catenanelike molecules and knotted rings[3-5] would be formed. In a somewhat bold conclusion Frisch, Martin, and Mark[6] explained the anomalous physical properties of polysiloxanes by postulating polymers composed of linear chains as well as chainlike bonded rings, each ring consisting in turn of 50 to 100 interlocked monomeric species. In addition, a variety of other rings were supposed to be present, ranging from very large to very small ones which could barely enclose a linear polysiloxane molecule. According to the authors, these chainlike molecules, the structure of which still remains to be proven, act as plasticizers and are responsible for the fact that polysiloxanes remain liquid to waxy in spite of their high molecular weights of 100,000 to 1,000,000.

From the depolymerization behavior of polymeric phosphonitrile chloride Patat and Derst believe it to be evident that it is built up from statistically interlocked rings.[7] The individual rings are supposed to be of low molecular weight (degree of polymerization from 10 to 50 trimers depending on the temperature), however, the interlocked aggregates possess a molecular weight of about 1,000,000. This explanation is supposed to account for the low solubility, the limited swelling of the main part of the polymeric phosphonitrile chloride in solvents, and the rubberlike elasticity of the polymer.

With the advances made in the synthesis of macrocyclic compounds, in the years between 1930 to 1950 (here the fundamental work of Ziegler *et al.*,[8,9] Hansley,[10,11] Prelog *et al.*,[12] and Stoll *et al.*,[13] should be mentioned) the thought of constructing catenanes became realistic.

The first experimental work on the synthesis of catenanes which took into consideration the actual ring size required and concerned itself with a realistic synthetic path was carried out at the University of Freiburg, Germany, by Lüttringhaus *et al.*[5] and was published in 1958. In 1960 Wasserman first reported the synthesis of a catenane.[1] He isolated "a few milligrams" of a substance, the structure of which has up to now not definitively been proven with mass spectrometry.[14] After years of intensive model investigations a successful directed synthesis of a catenane was accomplished by Schill in the year 1964.[15-17] A first communication was published by Schill and Lüttringhaus.[18] The structure of the catenane was confirmed by a mass spectrum measured by Vetter and Schill.[19] Since this first success the research group led by Lüttringhaus has continued to work on statistical methods, while Schill and co-workers have continued the investigations on the directed syntheses of catenanes.

Based on the reactions worked out in the chemical system used for the directed synthesis, Lüttringhaus and Isele were able to synthesize a catenane through a statistical, but sterically influenced method in the year 1967.[20,21]

In the same year Wang and Schwartz synthesized a catenane through the statistical method by cyclizing 5-bromouracil labeled coliphage λ DNA in the presence of cyclic phage DNA molecules. Theoretical and stoichiometric considerations about the formation probability of catenanes were confirmed in this work.[22]

The conjectures about the occurrence of catenanes in nature were recently confirmed by Vinograd *et al.*[23,24] They convincingly demonstrated by electron microscope examination the existence of interlocked closed circular duplex mitochondrial DNA in HeLa cells as well as in human leukemic leukocytes.

A system which has the same underlying principle as a catenane and for which the name rotaxane (Latin: *rota* = the wheel, *axis* = the axle; the ring is like a wheel on an axle. The dumbbell shape of the latter prevents the wheel from slipping off) has been suggested, can be represented by **2**.[25,26] In compounds of this class, bulky end groups prevent the extrusion of a threaded chain from a macrocycle. The existence of such aggregates is possibly due to the limited flexibility, size, and compressibility of its individual components.

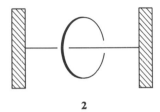

2

The synthesis of rotaxanes has been worked out cursorily by Freudenberg and Cramer[27] and somewhat more in depth by Stetter and Lihotzky.[28] Following the method of a directed synthesis, a rotaxane was first synthesized by Schill and Zollenkopf.[25,26] The synthesis of a rotaxane by a statistical method was reported by Harrison and Harrison in a communication.[29]

3

The synthesis of a trefoil knot **3** has up to now not been reported, although speculations have been made about the probability of its formation in polymerization and cyclization reactions.[3] First systematic work toward the synthesis of a knot is now in progress.[30-33]

The synthesis of catenanes was briefly summarized by Lüttringhaus.[34] A survey on the chemistry of rotaxanes and knots has not been published.

2

Chemical
Topology

A compound can generally be described,[3] unequivocally, by (1) the order in which given numbers of atoms are joined, (2) the type of bonds which connect them, (3) the configuration at asymmetric atoms or rigid centers, (4) the conformation, and (5) the topology.

Chemical topology deals with the structure and the property differences of compounds which are identical with regard to points (1), (2), and (3) and which in spite of that can not be interconverted by conformational changes, such as rotation about an axis or modification of bond angles. Interconversion can only take place by breaking and reforming chemical bonds. Frisch and Wasserman called compounds which can be classified in this manner topological isomers.[3] According to this nomenclature, the knot **3** and the macrocycle **4**, or the catenane **1** and the twice threaded catenane **6** are topological isomers. According to the suggestions of the authors, however, catenane **1** and the two rings **5** are also topological isomers even though the concept of isomerism should be restricted to chemical units with the same molecular formula.[35] If this restriction is ignored, as in the case of **1** and **5**, the sum of two separate entities would be isomeric to another entity consisting of their combination. For example, an inclusion compound would then be isomeric with its components. For this same reason, a rotaxane is not isomeric with its molecular subunits.

A catenane is composed of two distinct entities, and it is therefore a matter of definition whether it is considered as one or as two molecules. From the properties so far studied, however, especially the molecular weight determination, a catenane can undoubtedly be considered as one molecule.[16, 19] The stability of catenane systems is dependent, the same as in other chemical compounds, upon the weakest bond in either of the two rings. Since there is no chemical bond between the two rings, Frisch *et al.*[6] have called such a linkage a *mechanical* bond; Frisch and Wasserman[3] suggested the name *topological* bond.

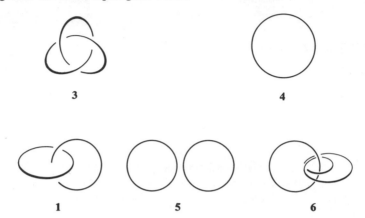

3

Nomenclature of Catenanes and Rotaxanes

Explicit suggestions for the naming of catenanes and rotaxanes have not been made so far. Frisch and Wasserman use the designation 34,34-catenane for a compound consisting of two 34-membered macrocycles.[3] In an article about absolute configuration and chemical topology, Tauber uses a similar terminology and names compound **7** as [34,34]-catenane-1,18,1′,18′-tetraone-1,1′-dioxime, a designation which easily leads to confusion.[36] Kohler and Dieterich in their interpretative work, which is not yet experimentally confirmed, suggest the word *cum* (Latin = with) be inserted between the individual components of a catenane. Hence, the catenane **7** would be named "cyclotetratriacontane-1,18-dione-1-oxime cum cyclotetratriacontane-1,18-dione-1-oxime."[37] According to the designation thus far used by Schill, compound **7** would be named "catenane from cyclotetratriacontane-1,18-dione-1-oxime and cyclotetratriacontane-1,18-dione-1-oxime."[16] Rotaxanes were named in an analogous manner.[25]

Since one can expect the synthesis of additional compounds with mechanical linkages in the near future, we suggest the following nomenclature. The compound concerned is designated as a catenane or rotaxane, this term appearing at the end of the name. The number of molecular sub-entities is given in brackets at the beginning of the name. Thus, the simplest catenane is a [2]-catenane. The brackets are followed by the names of the molecular entities which are enclosed in brackets as well. The rings of a catenane are numbered analogously to a normal or branched paraffin; the

$$N\text{-}OH$$
$$\|$$
$$C$$

$(CH_2)_{16}$

$(CH_2)_{16}$ $O{=}C$ $(CH_2)_{16}$ $C{=}N\text{-}OH$

$(CH_2)_{16}$

$$C$$
$$\|$$
$$O \qquad 7$$

number of each being placed ahead of its name. In the case of unbranched species such a numbering system is not necessary. The names of the rings forming a branch to the main chain are numbered with subscripts and placed together with the ring on which the branching takes place in one bracket.

In catenanes and rotaxanes containing multiple windings it is also necessary to designate the winding number α.[23] The quantity α represents the number of times one macrocycle winds about the other. Catenanes which form a ring with themselves like **14** are designated as cyclocatenanes.

To clarify the proposed nomenclature a few examples are shown (see structures **8–14**).

$(CH_2)_{20}$ $(CH_2)_{20}$

8

[2]-[cycloeicosane]-[cycloeicosane]-catenane

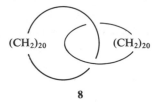

R——— $\text{-}(CH_2)_{10}$ ———R
$(CH_2)_{20}$
R = aryl

9

[2]-[1,10-diaryldecane]-[cycloeicosane]-rotaxane

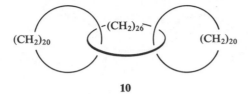

10

[3]-[cycloeicosane]-[cyclohexacosane]-[cycloeicosane]-catenane

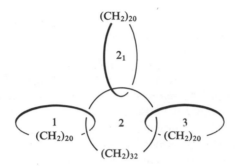

11

[4]-1-[cycloeicosane]-2-[cyclodotriacontane-2_1-cycloeicosane]-3-
[cycloeicosane]-catenane

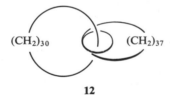

12

[2]-[cyclotriacontane]-[cycloheptatriacontane]-catenane   ($\alpha = 2$)

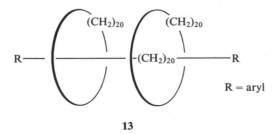

R = aryl

13

[3]-[1,20-diaryl-eicosane]-[cycloeicosane]-[cycloeicosane]-rotaxane

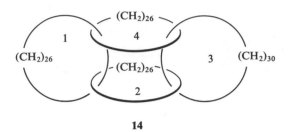

14

[4]-1-[cyclohexacosane]-2-[cyclohexacosane]-3-[cyclotriacontane]-4-
[cyclohexacosane]-cyclocatenane

As the chemistry of molecules with mechanical bonds is extended and with it the isomeric possibilities increase, further differentiations will have to be introduced into the nomenclature. For certain considerations the topological bonding number (TBN) introduced by Hudson and Vinograd is a useful concept.[23] The topological bonding number of a certain ring gives the quantity of rings which has to be opened to release this intact ring. In Fig. 1 a few examples are given. This designation, of course, does not specify the topological winding number α, nor does it give any information about the isomeric possibilities.

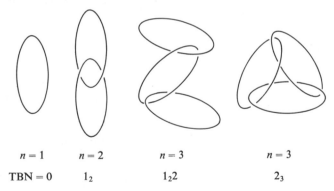

$n = 1$	$n = 2$	$n = 3$	$n = 3$
TBN $= 0$	1_2	$1_2 2$	2_3

FIG. 1.   Topological bonding number (TBN) isomers; n gives the number of rings in the corresponding catenane. The subscripts indicate the number of rings in the oligomer with a given topological bonding number (taken from Hudson and Vinograd[23]).

4

Stereochemistry of Catenanes, Rotaxanes, and Knots

The material so far published on the stereochemistry of catenanes and knots has been limited to relatively simple cases, while the stereochemistry of rotaxanes has not been dealt with at all. Neither have there been any publications on the experimental separation of enantiomeric catenanes or rotaxanes into their antipodes.

4.1. Catenanes

The stereochemistry of the simpler carbocyclic [2]-catenanes can be categorized as follows:

1. Catenanes bearing no substituents do not occur as antipodes.

2. Catenanes which have substituents in only one of the two rings are stereochemically related to the corresponding uncatenated macrocycles.

3. Catenanes in which each subunit contains two substituents located on the same ring atom, are stereochemically related to allenes and spiro compounds.[38] If each ring has two substituents A and B, as shown in formula **15**, a necessary and sufficient condition for the occurrence of enantiomers is that A $\neq$ B.[39]

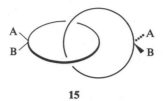

15

4. Catenanes where each subunit contains two substituents located on different ring atoms, generally exist as an enantiomeric pair.[40] Doornbos pointed out, however, that catenanes with two substituents do not always have to lead to enantiomers.[38] Such a case is illustrated in formula **16**. In catenane **16** each ring has the same two substituents R on different ring atoms. The segments of the individual rings of the catenane ($n \neq m$) and the configuration of the substituents have to be of such a nature that the molecule possesses a fourfold alternating axis of symmetry. In this case the catenane is achiral, although the component rings are dissymmetrical.

In other special cases, however, each ring of a catenane may possess a plane of symmetry, the catenane as a whole remaining chiral. Such a catenane is illustrated by formula **17**, in which one ring has substituents A and B, the other ring substituents C and D ($A \neq B$; $C \neq D$). The rings ($n = m$ or $n \neq m$) and the substituents have to be of such a nature that each ring possesses a plane of symmetry. Since the symmetry planes of the two rings intersect, the catenane is chiral.

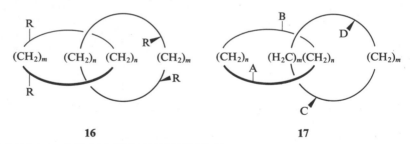

16   **17**

5. A type of enantiomerism closely related to catenane **15** was first pointed out by Closson,[40] and in more generalized form by Prelog *et al.*[40a] and later by Cruse.[41]

Catenanes **18** and **19** only differ by the reversal of the segment sequence in one ring. This difference makes them cycloenantiomers. In order that this difference be pertinent, each ring of a catenane must consist of at least three different segments, even though every segment may appear in both rings.

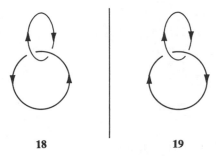

18   **19**

6. Special cases may arise if one of the two rings of a catenane is so small, or has substituents so large, that free rotation of one ring within the other is hindered.[38] Such a case is illustrated by formula **20**, where two enantiomers arise because the large substituents A and B prevent free rotation of the rings.

Another type of stereoisomerism may arise in catenane **20**, if not only $A \neq B$ but $n \neq m$ as well. In this case, two stereoisomers may exist as illustrated by catenanes **21** and **22**. Both stereoisomers should be separable into antipodes.

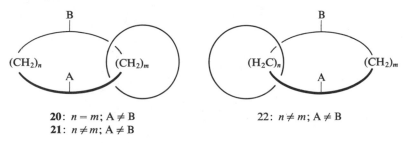

20: $n = m$; $A \neq B$ **22**: $n \neq m$; $A \neq B$
21: $n \neq m$; $A \neq B$

7. The [2]-catenane **6**, having a winding number $\alpha = 2$, is not identical with its mirror image and should therefore be optically active. This isomerism corresponds to the mirror image relation between an α- and a β-helix.

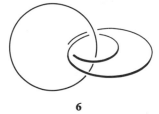

6

The stereochemistry of [3]-catenanes, as well as of higher catenanes, has not been considered so far except for some special cases.

A possibility to differentiate the [3]-catenanes **23** and **24** by means of optically active compounds, was pointed out by Frisch and Wasserman.[3]

23   **24**

It is interesting to note that the Borromean rings as pictured by **25**, or in a different view by **26**, are identical with their mirror images. This is true regardless of the direction of the ring sequence or other structure differences among the rings. It follows that compound **27** is identical with its mirror image.[36]

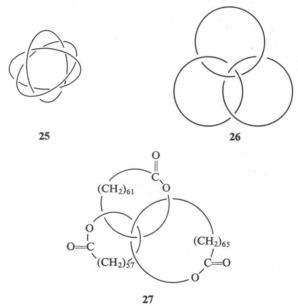

25   **26**

27

4.2. Rotaxanes

The stereochemistry of rotaxanes closely resembles that of catenanes, as is illustrated with catenane **20**. Even though rotaxane **28**, in which R ≠ R' and A ≠ B is achiral, it exhibits geometric isomerism.

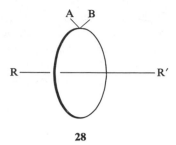

28

As in the case of catenanes, rotaxanes may exist as cycloenantiomers. Compounds **29** and **30**, which are mirror images, only differ in the direction of the ring segment sequences.

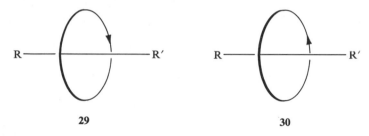

29   **30**

4.3. Knots

Knots have been treated mathematically by Tait,[42] and more recently by Reidemeister.[43] The simplest example of a knot is illustrated by the trefoil knot **3**. This compound is chiral and should be separable into optical antipodes.[43, 44]

3

A calculation of the magnitude of the optical activity of a trefoil knot was carried out by Kornegay *et al.*[45] according to the theory of Kirkwood.[46]

Likewise, knots of higher complexity can exist as mirror images which are not superimposable.[36] There are cases, however, in which they are optically inactive, i.e., the mirror images are superimposable. In most of the latter cases the knots become enantiomeric if an inner sequence direction exists. This then is another case of cycloenantiomerism.

5

Absolute Configuration

A suggestion for the determination of the absolute configuration of catenanes by extension of the Cahn-Ingold-Prelog rules[47-49] has been made by
Tauber.[36] The main feature of Tauber's proposal can be described as
follows[49]: One chooses in each ring the atoms with highest and second
highest order. If these atoms, which can be called, respectively, a and b for
one ring, and c and d for the other ring, are placed in pairs ab and cd as far
apart from each other as possible and then fitted to the tetrahedron **31**, the
secondary chirality of any pair of interlocked rings can be specified. If the
pairs ab and cd are identical, tetrahedron **32** can be used. In both instances

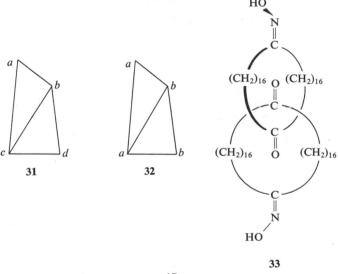

31

32

33

17

the chirality rules and standard subrules are now applied in the same manner as in the case of axial chirality. According to these rules, catenane **33** has the (S)-form.[36]

Although no suggestions have been made for the determination of the absolute configuration of rotaxanes it seems expedient to use a similar procedure as with the catenanes.

The determination of absolute configuration of knots is based on a normalized knot projection.[36] For every knot there exists a planar projection with a minimum of overcrosses or "double points" as they are called. If the knot is given an arbitrary direction then the upper segment at a double point has to be turned either clockwise or counterclockwise through an angle of less than 180° in order to align it with the lower segment. To each double point a characteristic ϵ can be assigned: $\epsilon = +1$, if the turn was clockwise, $\epsilon = -1$, if the turn was counterclockwise. The absolute configuration is then to be assigned as follows:

(R) if $\sum \epsilon > 0$

(S) if $\sum \epsilon < 0$

According to these rules knot **34** has the (R)-form ($\sum \epsilon = +3$).

34

Tauber was able to show that some knots for which $\epsilon = 0$ are identical with their mirror images regardless of the direction attributed to them. In other cases, knots for which $\epsilon = 0$ are identical with their mirror images if no direction is attributed to them.

6

Naturally Occurring Catenanes

Circular base-paired duplex DNA's are widely distributed under living organisms.[50, 50a, 50b, 51] Molecules of this type were first found in tumor polyoma viruses by Dulbecco and Vogt[52] and by Weil and Vinograd[53] in 1963. All these DNA's exist as covalently closed circular duplexes without chain ends.[52-54]

Closed circular duplex DNA's are double stranded and, hence, form two closed rings interlocked by a topological bond having a multiple winding number.[23, 55] Such two interlocked rings form if a linear double stranded DNA cyclizes with an even number of half twists, as can be demonstrated with a Möbius strip cut lengthwise.

The bond which mechanically links the catenated rings is characterized by a winding number α of about 450 in polyoma[55, 56] and SV40 tumor virus and of about 1500 in circular mitochondrial DNA.[23] The topological winding number α represents the number of times one strand winds about the other, when the molecule is placed in a plane.[23]

One may speculate,[57] that the duplex DNA not only consists of two interlocked rings but also to the same extent of a continuous unbroken knot composed of the two strands. This possibility, which would not appreciably change the structure of the circular duplex DNA as a whole, would arise when a linear double stranded DNA cyclizes with an odd number of half twists.

The physical and physicochemical properties of closed circular DNA differ in several aspects from those of linear or circular duplex DNA containing one or more single strand breaks or nicks.[55] The resistance to denaturation,[53, 58] the sedimentation velocity and the buoyant density in alkaline solution are enhanced in the closed circular molecules. All these properties can be explained by the inability of the polynucleotide strands to unwind.

The methods for the detection and the isolation of closed circular duplex DNA are based on the resistance to denaturation and the change of buoyant density in the presence of aminoacridines[59, 60] or ethidium bromide **35**.[61, 62]

35

which are capable of intercalating between DNA base pairs. The restricted uptake of the intercalating dye ethidium bromide forms the basis of a method for isolation and detection of closed circular DNA's.[23, 24, 50a] The binding of intercalating dyes causes a partial unwinding of the duplex structure of DNA molecules. It was concluded that the intercalation of one ethidium bromide molecule unwinds the duplex by 12°.[63] A closed circular DNA with no possibility to unwind resists the uptake of dye in high concentration and thus binds a smaller amount of dye as an open or nicked DNA. The binding of ethidium bromide, which has a low density, lowers the buoyant density of DNA in cesium chloride gradients. As a result, a mixture of open and closed DNA's forms two well-separated bands after centrifugation.

Hudson and Vinograd could show by band sedimentation combined with electron microscopy that mitochondrial DNA from HeLa cells contain catenated molecules.[23] The lowest band in cesium chloride gradient contains about 10% dimers and 90% monomers, while the intermediate band contains about 60% dimers and 40% monomers. From electron micrographs it is evident that the dimers in the lowest band consist of closed monomers which are catenated while the dimers from the intermediate band consist of one closed and one nicked circular, duplex DNA molecule. The doubly closed catenated dimer has the same sedimentation coefficient as the closed circular dimer.[64]

When the DNA molecules were measured in the electron microscope it was found that each one of the components of the catenated molecules measured approximately 5 μ, i.e., the same length as the unlinked monomers. Although other work[50a] already had mentioned the existence of oligomers with $n > 2$, the electron micrographs made the occurrence of catenated trimers, tetramers, pentamers, and even a heptamer readily visible.

In an analogous study by Clayton and Vinograd it was found that mitochondrial DNA of human leukemic leukocytes from two patients contained not only catenated dimers but also closed circular, duplex DNA molecules that were twice the length of a monomer. As shown, the lower band of one patient contained 26% circular dimer DNA from which 3% are catenated dimers. These findings are very interesting in view that, in normal cases, one finds no circular dimers and 3% catenated dimers.[24] Here then, the conjecture probably often talked about, but rarely mentioned in the literature—that catenanes were to be looked for in nature—has been confirmed.[65]

Other investigations of Vinograd *et al.* showed the presence of catenated mitochondrial DNA in 3T3 mouse cells transformed by SV40 virus and unfertilized eggs of sea urchins.[24, 50b] Additional experimental results suggest the existence of mechanically linked DNA's in other mitochondrial DNA.

Riou and Delain showed by electron microscopy that kinetoplastic DNA from *Trypanosoma cruzi* exists partly in the form of catenanes composed of two or more topologically interlocked circular units of monomer size 0.45μ.[51]

To gain a clear picture about the relation between the occurrence of oligomers and the physiological state of cells, further investigation is yet required.

To explain the occurrence of oligomer DNA's it was postulated that they arise either in the course of replication or during recombination. While the mode of replication is not yet understood, the formation of catenated oligomers may be explained by two mechanistic models. In one postulated model, which also explains the formation of dimers, the DNA circular duplex molecules first pair, then break and recombine again. This model also implies the formation of individual catenated species from circular dimers and circular monomers. This, however, has not been observed.[23]

According to another proposal made by Thomas, a four-stranded pairing region between the circular duplex molecules is not necessary for the formation of catenated dimers.[66] Instead, the molecules are specifically broken, matched with other broken molecules, and reunited. This model differs from the previous one in that catenated dimers can arise from monomers without the intermediate formation of cyclic dimers.

[2]-Catenanes

7

Methods for the Synthesis of Catenanes and Ring Size Requirements

The synthesis of catenanes is, as could be expected, only possible with macrocycles of suitable ring size. The inner diameter of one ring must be at least so large that it can accommodate the chain which will comprise the second ring. Fisher-Hirschfelder molecular models, which take into account the repulsive forces between atoms,[67] predict that macrocycles of at least 20 methylene groups are necessary to synthesize a carbocyclic [2]-catenane.[3] With Stuart-Briegleb molecular models, on the other hand, a [2]-catenane can already be built with macrocycles comprising 18 methylene groups. From molecular models, however, no definite conclusions can be drawn about the formation tendencies of catenanes having these ring sizes. As in the case of some cyclophane syntheses, catenanes of small ring sizes will most likely be obtained by ring-contraction reactions, even though side reactions between the molecular subunits may cause complications.

While considering the possibilities for a catenane synthesis, it is useful to distinguish between the case where catenanes are formed as by-products and the case where catenanes are the sole product of the reaction. The first approach will be designated as a *statistical* method and the second as a *directed* method.

7.1. Statistical Methods

In statistical methods a catenane and its molecular subunits are formed side by side in a definite proportion, as determined by statistical laws. The controlling factors are size and structure of the starting materials, conformational aspects, steric hindrance, etc. This means that in a given chemical system, a catenane is formed in a certain yield that may be varied with experimental conditions. Since the theory of conformational behavior of macrocycles and long-chain molecules in solution is as yet little understood, definite predictions can not be made about the formation probability of catenanes.

A series of theoretically plausible paths for the statistical synthesis of catenanes were proposed by Lüttringhaus et al.[5] All of these are based on the simple assumption that two interlocking rings may be formed from an open chain which is threaded through a ring.

The proposals of Lüttringhaus and co-workers may be classified into four schemes, to which a fifth one can be added[57]:

7.1.1. Scheme 1

The cyclization of a long chain **36** with terminal functional groups X in the presence of macrocycles **37** may result in catenanes if the chain **36** assumes

the intraannular conformation **38** at the time of the ring closure. The probability of catenane formation by this method is not very high, but may be increased through additional steric factors and through a temporary linkage of the macrocycle **37** to the long-chain compound **36**, as discussed in Scheme 2.

7.1.2. *Scheme 2*

The long-chain compound **36** is linked to **37** through an auxiliary link-age,* which is shown as a dotted line, to give **39** and **42**. In this manner the two molecules are forced to remain close to each other. By choosing an appropriate auxiliary linkage, it may be possible to influence the confor-mational equilibrium between **39** and **42**. Cyclization results in the linked rings **40** and the precatenane **43**. On cleaving the auxiliary linkages, they are converted to two macrocycles **41** and a catenane **1**, respectively.

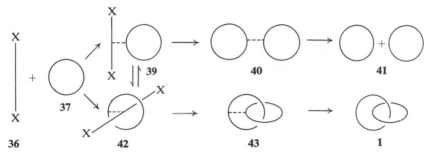

7.1.3. *Scheme 3*

An intraannularly bridged macrocycle **44** is linked, by way of its bridge, to one of the functional end groups of a long-chain compound **36**. Cycliza-tion of the resulting compound **45** should give the extraannularly bonded compound **46** and a precatenane **47**. The rupture of the auxiliary linkages gives catenane **1** and its molecular subunits.

* The term auxiliary linkage, denotes such chemical linkages which are necessary for the synthesis of the intermediates, but which are cleaved during the later stages of the synthesis (see Ziegler[68]).

7.1.4.   *Scheme 4*

The fourth scheme is closely related to Scheme 2. Two long-chain compounds **36** with functional end groups are linked to give **48**. Since the cyclization takes place in two steps, essentially the same conditions as in Scheme 2
are established, with the exception that an additional bridged macrocycle
49 is formed. On cleaving the auxiliary linkages in the latter, macrocycle **50**
is formed.

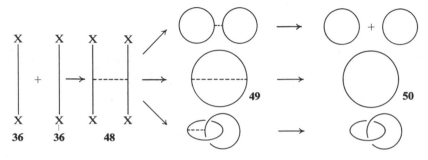

7.1.5.   *Scheme 5*

The schemes dealt with so far have in common (with the exception of
Scheme 3), both rings of the precatenane linked only once. If multiple
auxiliary linkages are used, the probability of catenane formation can be
substantially increased. A long-chain compound with functional end
groups Y and a macrocycle with the functional group X are linked to give
51 and **54**. As can be seen, the substituents can be located at different places.
The functional end groups Y should be of such a nature that they can react
with X but not among themselves. Hence, when the auxiliary linkages and
the bond between the substituent X and the macrocycle are broken in the
cyclization products **52**, **53**, **55**, **56**, catenane **1** and its subunits should be
among the products.

All the statistical methods mentioned have in common, molecular subunits that form alongside the catenane. Strictly speaking, only the first
method is truly statistical, since in the other cases an attempt has been made
to influence the statistics by bringing about favorable steric conditions.
For this reason, Lüttringhaus and Isele coined the term "semistatistical"
for methods in which it was attempted to influence the probability of cate-

nane formation by sterically influencing the ring closure method with the choice of the components used.[20]

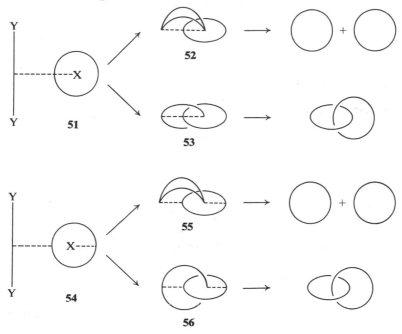

7.2.   Directed Syntheses

The directed synthesis consists in bringing about such a steric arrange-ment between the components, which will form the subunits of the cate-nane, that the intraannular cyclization toward a precatenane is the only ring closure possible. The path for such a synthesis, as first pointed out by Schill,[15, 69, 70] is pictured in Fig. 2.

A planar ring **59** carrying substituents A and D as well as two chains R is combined with chain **60** carrying the functional group B in such a way that in the resulting compound **61** the chain is fixed perpendicularly to the plane of the ring at the linkage point. The nature of the functional end groups C is such that they can react with D but not with each other. Cyclization of **61** leads to the intraannularly linked diansa compound **62**; due to the per-pendicular linkage in **61** and a chain length which is kept as short as possible

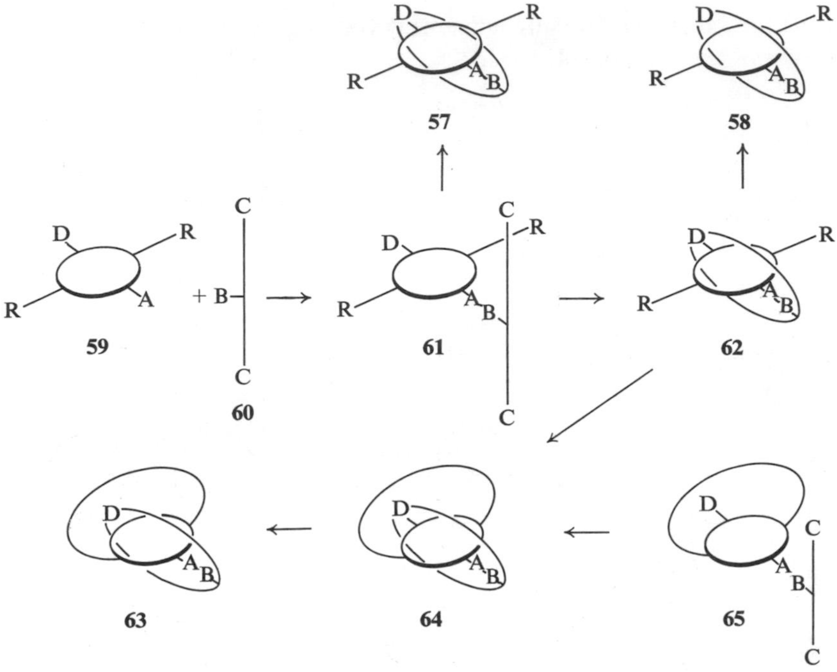

FIG. 2.   Schematic representation of a directed synthesis of catenanes (reprinted from Schill[69] with permission of the publisher).

the extraannular compound **57** can not be formed. Whether the perpendicularity at the linkage point between chain and planar ring is necessary in all cases remains questionable. In most cases the space taken up by the ansa bridges make the formation of symmetrical diansa compounds more likely in any case.

Cyclization of the substrates R in **62** results in the precatenane **64**. By detaching the planar ring from the macrocycle formed by the two diansa bridges and substituent D, catenane **63** is obtained. A second possibility, synthesizing precatenane **64**, which differs only in the sequence of the reaction steps, consists in cyclizing **65**.

By bringing about the proper conditions and fulfilling certain requirements, it should be possible to convert all the statistical methods schematically illustrated above to methods for a directed synthesis of catenanes. In a sense, the directed synthesis mentioned may be regarded as a special case of the fifth scheme discussed in the statistical syntheses of catenanes.

If both substrates R in **62** are bulky groups, a rotaxane **58** is obtained when the bonds between A and B and between the planar ring and D are cleaved.

7.3. Möbius Strips

Another general method for the synthesis of catenanes and knots depends on the principle of the Möbius strip.[3, 71-73] The "endless, edgeless and one-sided" Möbius strip is formed by twisting for example a strip of paper by half a turn and joining the ends.

Possibilities for the synthesis of catenanes and knots can be seen when circular strips with n half twists are cut lengthwise into halves, thirds, quarters, etc. The results are different in each case and, at first view, surprising.

A strip with an even number n of half twists always gives, when halved, two individual strips related in various ways to each other. For $n = 0$ one obtains two separated rings, for $n = 2$ a catenane results, for $n = 4$ a catenane with winding number $\alpha = 2$ results, for $n = 6$ a catenane with $\alpha = 3$ is produced, hence $\alpha = n/2$. These results are shown in Figs. 3 and 4.

A strip with an odd number of half twists, when cut lengthwise into halves, always gives single rings, knotted in various ways. For example,

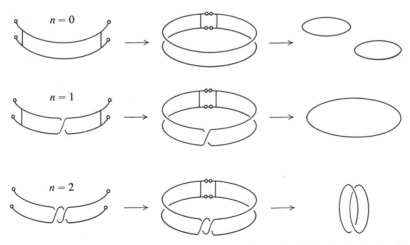

FIG. 3.   Configurations resulting from the cyclization of n half-twisted 2-strips after cross-link scission (taken from Van Gulick[72]).

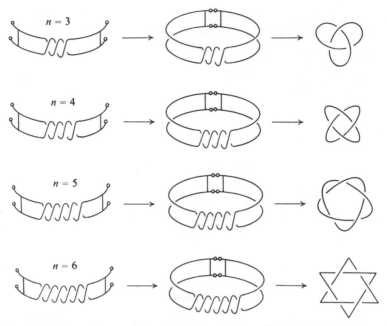

FIG. 4.   Configurations resulting from the cyclization of *n* half-twisted 2-strips after cross-link scission (taken from van Gulick[72]).

for $n = 1$ a ring is obtained twice as large but one-half as thick as the original ring, for $n = 3$ a trefoil results, and for $n = 5$ a cinquefoil is obtained. In other words, for all odd half twists, except $n = 1$, n equals the number of crossings in the knot obtained. The rings and knots obtained from Möbius strips are in turn composed of twisted bands but these details can not be discussed here.

The results from trisecting a Möbius strip of n half twists is shown in Fig. 5. For $n = 1$, a catenane is obtained having one link the same diameter as the original strip and one link with twice this diameter. If $n = 2$ three

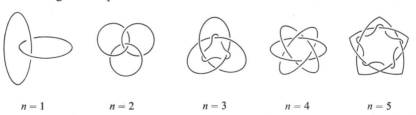

| $n = 1$ | $n = 2$ | $n = 3$ | $n = 4$ | $n = 5$ |

FIG. 5.   Configurations resulting from cyclization of *n* half-twisted 3-strips after cross-link scission (taken from van Gulick[72]).

interlocking rings are formed, for $n = 3$ a trefoil is formed which has an additional ring intertwined. One can recognize that when n is even three variously interlocked rings are formed and when n is odd, except for $n = 1$, an n-foiled knot is obtained which has an additional ring wound through it.

A synthesis of catenanes or knots with the aid of Möbius strips could be statistical or directed, depending on the reaction path chosen. No publications have appeared so far in which this principle was applied to a chemical system.*

* After completion of this manuscript, R. Wolovsky [*J. Amer. Chem. Soc.* **92**, 2132 (1970)] and D. A. Ben-Efraim, C. Batich, and E. Wasserman [*J. Amer. Chem. Soc.* **92**, 2133 (1970)] gave mass spectrometric evidence for the formation of [2]-catenanes via a "Möbius-strip" approach in the metathesis reaction of cyclododecene.

8

Investigations of the Statistical Syntheses of Catenanes

8.1. Scheme 1

Several investigations have shown, that polymers contain a certain portion of macrocycles.[4] It can therefore be expected that catenanes are present as well, although their formation probability is small. No attempts to isolate catenanes have been reported so far. This is understandable, since one can expect only very small differences in the physical properties of these and macrocycles of equal molecular weight and sequence. Knowing, however, some properties of catenanes and being able to definitively identify them with mass spectrometry, it should be possible to isolate them from appropriate low molecular fractions.

The simplest route to obtain catenanes, the cyclization of compounds in the presence of macrocycles, is also experimentally the easiest one to carry out. A priori, however, one can also expect the lowest yields with this method.

Syntheses which fall into Scheme 1 of our classification have been experimentally investigated by Wasserman[1] as well as by Wang and Schwartz.[22]

In the work of Wasserman the starting material diethyl tetratriacontane-dioate 66, was cyclized to the acyloin 67 by the method of Hansley,[11] Prelog et al.,[12] and Stoll et al.[13] This in turn was reduced with deuterated hydrochloric acid by the Clemmensen method thus obtaining in 50% yield the cyclotetratriacontane **68**, containing an average of five deuterium atoms per molecule. The diester **66** was now cyclized in a 1:1 mixture of xylene and the deuterated cycloparaffin **68**, the latter being in a 1 hundredfold excess with respect to **66**.[74] The acyloin was obtained in 5–20% yield. To prove the formation of a catenane, the reaction mixture was freed from the unchanged deuterated cycloparaffin **68** through column chromatography. The acyloin fraction was then oxidized with hydrogen peroxide in an alkaline solution whereby the dicarboxylic acid **69** was obtained. The presence of the deuterated cycloparaffin **68** in the products of this oxidation showed that the catenane **70** has been formed. The amount of the liberated cycloparaffin **68** was roughly proportional to that of **67**. With the reaction conditions mentioned, the yield of catenane **70** was 1% based on the acyloin **67** which means that starting with 10 g of cycloparaffin **68** 1 mg of catenane would be obtained. According to the author, the yield of pure catenane based on the starting material sebacic acid is 0.0001%.[14, 74] He states that "a few milligrams" of the catenane **70** have been isolated by this method. Unfortunately, no unequivocal structure proof has been given so far (for example, a mass spectrum).

$$C_2H_5OOC-(CH_2)_{32}-COOC_2H_5 \longrightarrow$$

66

(CH₂)₃₂   C=O   HC—OH

67

$$\longrightarrow \quad C_{34}H_{63}D_5$$

68

$$HOOC-(CH_2)_{32}-COOH$$

69

$C_{34}H_{63}D_5$ (CH₂)₃₂   C=O   HC—OH

70

Wang and Schwartz synthesized noncovalent closed catenanes by the cyclization of coliphage λDNA in the presence of hydrogen-bonded circular 186 phage DNA molecules.[22]

The DNA of coliphage λ has cohesive sites at the ends which are short segments of single-stranded polynucleotides with base sequences complementary to each other.[75-77] At low concentration they form circular hydrogen-bonded monomers when two sites of the same molecule cohere.[75] At high concentrations they form dimers and higher oligomers. According to Baldwin *et al.* the coliphage 186 DNA also possesses cohesive ends which are similar but not identical to those of the λ phage.[78]

A catenane was synthesized as follows. DNA isolated from phage 186[78] was cyclized obtaining 58 weight per cent monomer and 21 weight percent dimer as measured by band sedimentation. Next 5 μg/ml of 5-bromouracil labeled λcI_{857} DNA, prepared according to the method of Baldwin *et al.*,[78] were mixed with various amounts of the cyclic 186 DNA and then annealed. The formation of a catenane was indicated by the appearance of a new species of intermediate buoyant density when the reaction products were centrifuged in cesium chloride. Results are shown in Table 1, whereby

$$K_{dc} = \frac{[C_1^{186}:C_1^{\lambda}]}{[C_1^{186}][C_1^{\lambda}]}$$

TABLE 1

FORMATION OF CATENANES BETWEEN 186 DNA AND λcI_{857} DNA[a]

$[C_1^{186}]$, molecules/cm³	$[C_1^{186}:C_1^{\lambda}]/[C_1^{\lambda}]$	K_{dc}, cm³
3.95×10^{11}	0.088	2.2×10^{-13}
3.95×10^{11}	0.074	1.9×10^{-13}
3.95×10^{11}	0.094	2.4×10^{-13}
1.98×10^{11}	0.043	2.2×10^{-13}
1.98×10^{11}	0.045	2.3×10^{-13}

[a] Taken from Wang and Schwartz[22].

From the table it can be calculated that the yield of catenane is doubled from about 4.4% to about 8.5% when the concentration of cyclic 186 DNA is doubled from 11 μg/ml to 22 μg/ml.

Beside the new band already mentioned, a second smaller peak was observed corresponding to a trimer species composed of two 186 DNA and one λcI_{857} phage DNA. The authors believe that since 21% of the cyclic

186 DNA is present as a dimer this additional peak is caused by catenanes composed of the latter and the cyclic λ phage DNA.

8.1.1. *Probability Calculations for the Formation of Catenanes*

The application of statistical laws to the formation probability of catenanes has been discussed by Frisch and Wasserman[3] for the cyclization of hydrocarbon chains and by Wang and Schwartz[22] for the cyclization of DNA chains.

Frisch and Wasserman assume in their calculations that the rings and chains are present in solution as statistical coils to which certain radii R_1 and R_2 can be assigned. If in a large volume V, a relatively small number N_2 of linear molecules is cyclized in the presence of a relatively large number N_1 of cyclic molecules, the probability p_{12} for catenane formation is $\bar{\beta}$ times the probability of overlap of their segment distribution[3]:

$$p_{12} = \bar{\beta} \tfrac{4}{3}\pi (R_1 + R_2)^3 / V$$

$\bar{\beta}$ is a constant which accounts for the fact that overlapping does not necessarily lead to catenane formation. From trials with simple models, the magnitude $\bar{\beta}$ was estimated to be about $\tfrac{1}{2}$.

In a first approximation, a methylene group can be regarded as a segment. To calculate the yield of catenane, a relation is introduced between the radii and the segment numbers.[79] In addition, a correction factor is used to compensate for the volume of the ring segments in rings with a moderate number of segments. Carrying out the calculations for a catenane consisting of two 34-membered rings, a yield of 4% is obtained. More than an agreement of magnitude can not be expected in these calculations since the length of a methylene segment can only be estimated. The length of a segment is 1.54 Å, but due to the tetrahedral geometry of the carbon atom it is diminished to 1.26 Å in a methylene chain. For the above calculations an estimated value of 1.40 Å was used.[3]

An expression similar to that of Frisch and Wasserman without the geometric factor $\bar{\beta}$ was derived by Wang and Schwartz[22] for the equilibrium constant K_{dc} of the reaction

$$\underset{\text{cyclic 186 DNA}}{C_1^{186}} \quad + \quad \underset{\text{cyclic } \lambda \text{ DNA}}{C_1^{\lambda}} \quad \overset{K_{dc}}{\underset{}{\rightleftharpoons}} \quad \underset{\text{catenane}}{C_1^{186}\!:\!C_1^{\lambda}}$$

$$K_{dc} = \frac{[C_1^{186}\!:\!C_1^{\lambda}]}{[C_1^{186}][C_1^{\lambda}]} = \tfrac{4}{3}\pi(R_{c1} + R_{c2})^3$$

where R_{c1} and R_{c2} are the root mean square radii of two cyclic DNA molecules. The calculation of the equilibrium constant with this expression gives a value which is in reasonably good agreement with average experimental results.

8.2. Scheme 2

Investigations falling under Scheme 2 in our classification have been carried out by Lüttringhaus in co-work with Cramer, Prinzbach, and Henglein,[5] with Preugschas,[80] with Schill,[70, 81] with Vollrath,[82−84] and with Isele.[20, 21, 85]

In order to obtain, upon cyclization, two intraannularly joined rings, an intraannular conformation must exist between the chain to be cyclized and the macrocycle attached to it. To favor this conformation, the above mentioned authors used macrocycles composed of 21–28 links. The relatively stable inclusion compound between the cyclic trimer of 6-aminocaproic acid and benzene, observed by Zahn and Determann,[86] led to the assumption that a macrocycle which allows only a small amount of clearance for the chain within it also retains this chain to a certain extent. According to Stuart Briegleb molecular models this condition should be fulfilled with a 22- or 23-membered ring. A 20-membered ring can scarcely be fitted around a methylene chain, a 21-membered ring can enclose such a chain more easily, a 22- or 23-membered ring permits easy passage of a methylene chain and yet remains in close contact with it.

Lüttringhaus, Cramer, Prinzbach, and Henglein attempted to obtain catenanes by cyclizing long-chain dithioles embedded as inclusion compounds in cyclodextrins.[5] This method was chosen because of the water solubility and chemical nature of the cyclodextrins.

Cyclodextrins have the property of forming inclusion compounds which do not dissociate completely even in solution.[87] Since benzene and simple derivatives thereof form especially stable inclusion compounds, the hydroquinone ethers **71** and **74** and the mono- and dichloro derivatives **72** and **73** were synthesized.

The inclusion compounds of the derivatives with α- and β-cyclodextrins contained 3–9 % dithioles. Cramer and Henglein found that the molar fraction of dithiole in the inclusion compound diminished as the chain length of the included chain increased.[88] This fact may indicate that the extended

HS—(CH₂)₁₀—O— (ring, R' at top, R' at bottom) —O—(CH₂)₁₀—SH

71: R = R′ = H
72: R = H, R′ = Cl
73: R = R′ = Cl

HS—(CH₂—CH₂—O)₃— (ring) —(O—CH₂—CH₂)₃—SH

74

long-chain compound lies embedded in a tube of cyclodextrin molecules. If the 1,4-oriented chains protrude far enough out of the cyclodextrin tube, an intramolecular cyclization of the dithiole should be possible.

Since the host molecule is hydrophobic on the inside and hydrophilic on the outside, oxygen atoms were introduced into the side chains as in **74**. The hydrophilic polyethylene ether chains were intended to favor the drawing out of the side chains into the water and prevent their being further enveloped by cyclodextrin molecules. The oxygen atoms in the side chains were also supposed to be slightly attracted to the hydroxyl groups on the periphery of the cyclodextrin molecule, thereby bringing about a favorable conformation for cyclization.

In preliminary investigations some of the synthesized dimercaptans were cyclized, without cyclodextrin, with air and a copper(II)–salt catalyst to the corresponding disulfides. These macrocyclic disulfides are unstable compounds which already polymerize in boiling benzene. In the case of **71** only 2–3% yield of macrocyclic disulfide was obtained.

When the inclusion compound of **71** with α-cyclodextrin was cyclized and the inclusion compound destroyed with hot water, only unchanged dithiole was obtained. No evidence of cyclodextrin disulfide adduct formation could be found. The fact that the dithiole was not attacked in the dehydrogenation is explained by assuming that the side chains are closely enveloped by the cyclodextrin molecules and thereby shielded from any reaction. Only results for the cyclization of **71** have been thus far described, but according to the authors, cyclization of other dithioles gave similar results.

Lüttringhaus and Preugschas attempted to obtain an aminal **77** from the diaza macrocycle **75** and the ketodicarboxylic ester **76**.[80] The reaction was unsuccessful under basic as well as acidic conditions; not even hydrogen bonds between the carbonyl and the amino groupings being observed.

$$
\begin{array}{ccc}
\underset{\textstyle 75}{\left[\begin{matrix} -(CH_2)_{10}- \\ HN \qquad\qquad NH \\ -(CH_2)_{10}- \end{matrix}\right]} \;+\;
\underset{\textstyle 76}{\begin{matrix} COOCH_3 \\ | \\ (CH_2)_{11} \\ | \\ O{=}C \\ | \\ (CH_2)_{11} \\ | \\ COOCH_3 \end{matrix}}
\quad\xrightarrow{\;\;\;\not|\;\;\;}\quad
\underset{\textstyle 77}{\begin{matrix} COOCH_3 \\ | \\ (CH_2)_{11} \\ \left[\begin{matrix} \qquad\quad (CH_2)_{10} \\ N{-}C{-}N \\ (CH_2)_{10}\qquad\quad \end{matrix}\right] \\ (CH_2)_{11} \\ | \\ COOCH_3 \end{matrix}}
\end{array}
$$

To obtain the macrocycle **75**, 1,10-bis-*p*-toluenesulfonamidodecane was cyclized with 1,10-dibromodecane in tetrahydrofuran/butanol in the presence of sodium butoxide. The cyclization was carried out with 38% yield by means of the dilution method according to Ziegler and Orth.[89] The cleavage of the sulfonamide substrates with hydrobromic acid could only be accomplished with moderate yield. Higher yield of the macrocycle **75** could be achieved after the method of Stetter and Marx[90] in which the cyclic diamide from 1,10-diaminodecane and sebacic acid chloride is reduced with LiAlH$_4$ in tetrahydrofuran.

The diester **76** was obtained by carrying out a malonic ester synthesis with 1,21-dibromoheneicosan-11-one.[15, 91] In additional experiments, the diester **76** was converted to the oxime **78** which was catalytically reduced with Raney nickel to the amine **79** and subsequently converted with phosgene to the isocyanate **80**. In order to synthesize a model compound which was to be reacted with the macrocyclic diaza compound **75** and then cyclized, the isocyanate **80** was reacted with diethylamine to give the urea derivative **81**. Since the cyclization of this diester with sodium in xylene[11–13] was unsuccessful, no further attempts were made to synthesize a urea derivative from **80** and the diaza macrocycle **75**.

Further attempts to synthesize catenanes by methods falling under Scheme 2 of our classification were undertaken by Lüttringhaus and Schill.[70, 81]

Through reduction with LiAlH$_4$, the diol **83** was prepared from the acyloin **82**.[92] Approximately equal amounts of the *meso* and *racemic* form were obtained. The *meso*-diol was acetylated and subjected to free radical addition of hydrogen bromide, thereby obtaining the dibromide **84**. The

$$\underset{\textbf{78}}{H_3COOC-(CH_2)_{11}-\overset{\displaystyle N\diagup^{OH}}{\overset{\|}{C}}-(CH_2)_{11}-COOCH_3}$$

$$\underset{\textbf{79}}{H_3COOC-(CH_2)_{11}-\overset{\displaystyle NH_2}{\overset{|}{C}H}-(CH_2)_{11}-COOCH_3}$$

$$\underset{\textbf{80}}{H_3COOC-(CH_2)_{11}-\overset{\displaystyle N=C=O}{\overset{|}{C}H}-(CH_2)_{11}-COOCH_3}$$

$$\underset{\textbf{81}}{H_3COOC-(CH_2)_{11}-\overset{\displaystyle NH-CO-N(C_2H_5)_2}{\overset{|}{C}H}-(CH_2)_{11}-COOCH_3}$$

dibromide, in turn, was reacted with potassium cyanide to **85**, with saponification of the ester groups occurring in this step.

When **83** was converted to the boric acid ester, radical initiated hydrogen bromide addition could be achieved as well. Saponification of dinitrile **85** and esterification of the dicarboxylic acid thus obtained resulted in the dimethyl ester **86**.

Following the method of Salmi,[93] compounds **85** and **86** were ketalized in high yield with cyclodocosanone to **87** and **88**, respectively. *p*-Toluenesulfonic acid was used as the catalyst. These experiments showed that the method of Salmi can be successfully extended to long-chain ketones, which are often not very reactive,[94] as well as long-chain 1,2-dihydroxy compounds.

A 5-membered ketal was chosen for the junction between the macrocycle and the bifunctional long chain because it allows a right angle orientation between the two components at the junction. This geometry was thought to favor the formation of an intraannular compound upon cyclization.

Although it is simpler in preparation to obtain a long-chain keto compound and a macrocycle containing vicinal hydroxy groups, the opposite arrangement was chosen because of the following reasons. Prelog, Wirth, and Ruzicka[95] expect that the conformation of macrocyclic ketones is influenced by two factors: (1) The tendency of the methylene chain to remain in the zig-zag form which is the state of lowest energy[96] and (2) the tendency of the carbonyl oxygen to form hydrogen bonds with protons at an ade-

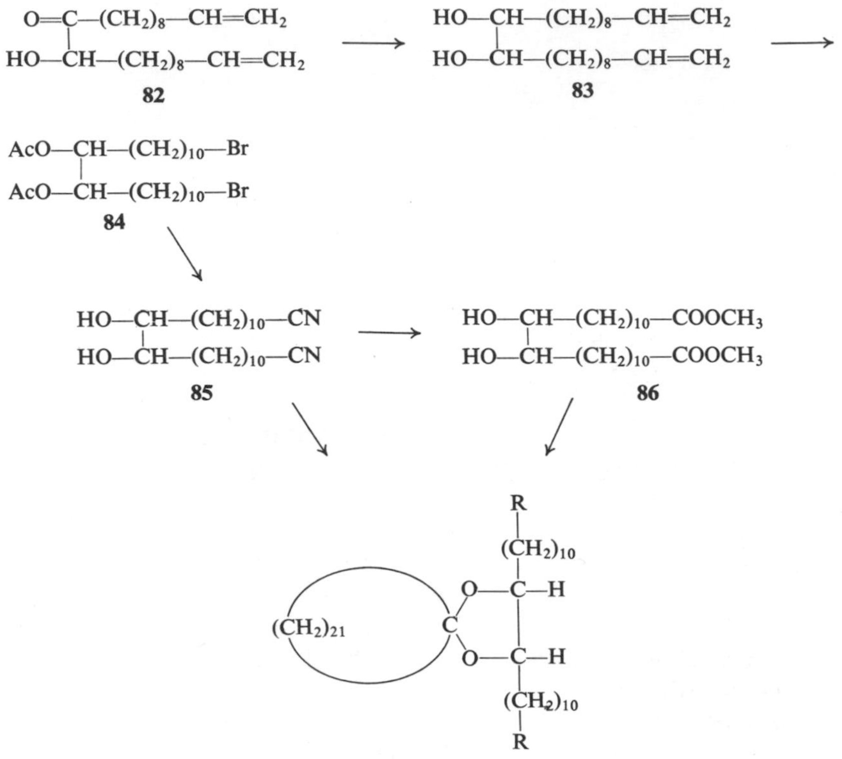

O=C—(CH$_2$)$_8$—CH=CH$_2$
|
HO—CH—(CH$_2$)$_8$—CH=CH$_2$

82

⟶

HO—CH—(CH$_2$)$_8$—CH=CH$_2$
|
HO—CH—(CH$_2$)$_8$—CH=CH$_2$

83

⟶

AcO—CH—(CH$_2$)$_{10}$—Br
|
AcO—CH—(CH$_2$)$_{10}$—Br

84

↓

HO—CH—(CH$_2$)$_{10}$—CN
|
HO—CH—(CH$_2$)$_{10}$—CN

85

⟶

HO—CH—(CH$_2$)$_{10}$—COOCH$_3$
|
HO—CH—(CH$_2$)$_{10}$—COOCH$_3$

86

↘   ↙

R
|
(CH$_2$)$_{10}$
|
O—C—H
|
(CH$_2$)$_{21}$ C
|
O—C—H
|
(CH$_2$)$_{10}$
|
R

87: R = CN
88: R = COOCH$_3$

quate distance. Evans et al.[97], in particular, argued for and postulated the existence of such hydrogen bonds. Considering the factors described above, one is led to a model for macrocyclic ketones in which the carbonyl oxygen is oriented toward the inside of the ring. In a number of publications Prelog, Ruzicka, and co-workers were able to show with classical methods of organic chemistry, that the carbonyl conformation changed with ring size.[98-101] On the basis of NMR measurements Ledaal[101a] also believed that the 11- and 12-membered cyclic ketones prefer a conformation with the carbonyl group oriented toward the inside of the ring. Since then Allinger and Maul showed that, contrary to the original postulate, at least in 8-membered cyclic ketones no hydrogen bonds are present.[102]

According to Prelog, a 15-membered cyclic ketone still favors a conformation in which the carbonyl is within the ring. It was therefore hoped that the

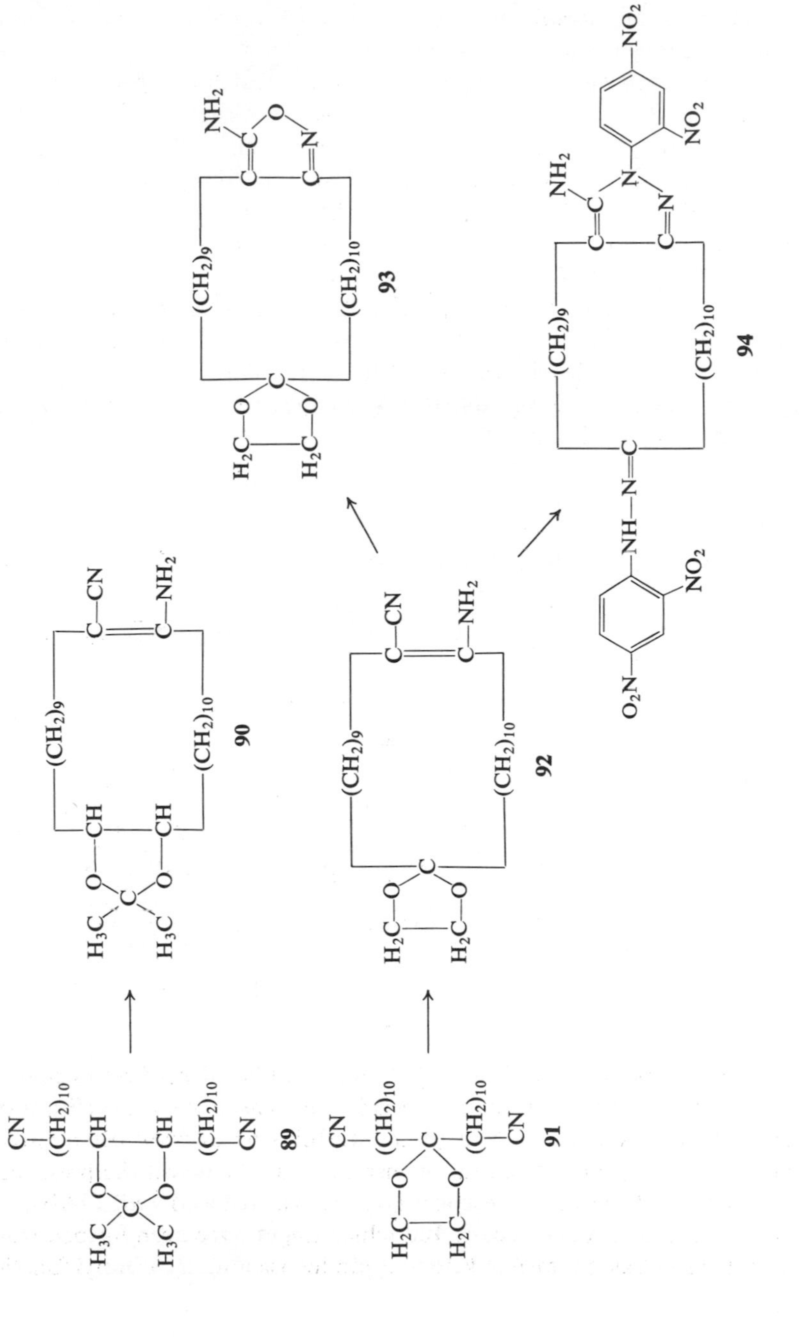
89
90
91
92
93
94

formation of an intraannular ketal was still favored when cyclic ketones with just enough inner room were used.

Since the cyclization method of Ziegler et al.[8] had not yet been carried out on dinitriles containing ketal functions, the two model compounds **89** and **91** were first cyclized. Using the Ruggli-Ziegler high-dilution method, the macrocyclic enaminonitriles **90** and **92** were isolated in 29 and 49% yields, respectively. When **92** was reacted with hydroxylamine, **93** was obtained, while the reaction with 2,4-dinitrophenylhydrazine in an acidic medium gave the derivative **94**.

The cyclization of ketal **87** resulted in an oil from which the enaminonitrile **95** could be isolated as a crystalline substance. The structure of the compound was proven by cleaving the ketal and obtaining the starting ketone cyclodocosanone and the dihydroxy compound **96**. No other reaction products could be isolated.[70]

$$
(CH_2)_{21}\ C\ \begin{cases} O\!-\!C\!-\!H \\ O\!-\!C\!-\!H \end{cases}\ \overset{\displaystyle (CH_2)_9}{\underset{\displaystyle (CH_2)_{10}}{}}\ \begin{cases} C\!-\!CN \\ C\!-\!NH_2 \end{cases}
$$

95

$$
(CH_2)_{21}\ C\!=\!O\ +\ \begin{matrix} HO\!-\!C\!-\!H \\ HO\!-\!C\!-\!H \end{matrix}\ \overset{\displaystyle (CH_2)_9}{\underset{\displaystyle (CH_2)_{10}}{}}\ \begin{matrix} HC\!-\!CN \\ C\!=\!O \end{matrix}
$$

96

Stoll, Hulstkamp, and Rouvé[103] showed that ketalized diesters could be cyclized to macrocyclic acyloins in good yield. Therefore, the cyclization of ketal **88** was expected to lead to the acyloins **97** and **99**, depending on whether the bonding was extra- or intraannular. To detect the presence of any catenane formed, the reaction mixture was reduced with LiAlH$_4$ and then hydrolyzed. Any catenane **100** which might have been formed would have, the same as the starting ketone cyclodocosanone, a carbonyl function.

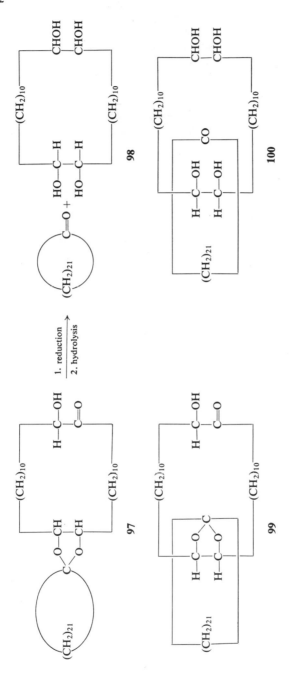

After the separation of the macrocyclic tetrahydroxy compound **98**, which probably consists of a mixture of stereoisomers, the mother liquor was treated with 2,4-dinitrophenylhydrazine. After separating the crystalline cyclodocosanone 2,4-dinitrophenylhydrazone, the mother liquor was analyzed by paper chromatography; no other hydrazones could be found, however. The catenane **100**, with its free keto group, should have formed a dinitrophenylhydrazone which in turn should have been distinguishable on a paper chromatogram. It was assumed that the keto group in compound **100** is still capable of reacting. This is reasonable, since, by conformational changes, the carbonyl group can turn to the outside and thus be in a position to react. Consequently, it had been shown that catenane **100** and its predecessor **99** did not form at all, or formed only in amounts not detectable by the method of analysis used.

Another reaction sequence which fits into Scheme 2 of our classification was carried out by Lüttringhaus and Vollrath.[82, 83] They planned to synthesize the ethers **103** and **104** from a macrocyclic nitrophenol **101** and a long-chain iodide **102** containing cyclizable end groups. The expected mixture of intra- and extraannular bonded ethers was then to be cyclized. Following a procedure given by Prelog et al.,[99a, 104] the nitrophenol **101** was synthesized in 89% yield by condensing cyclotetracosanone with nitromalonic dialdehyde in 90% alcohol and using ethanolic potassium hydroxide as a base.

Since the nitro group would probably interfere with the dinitrile cyclization, it was replaced on a model compound with hydrogen. For this purpose heneicosa-1,20-dien-11-one and nitromalonic dialdehyde were reacted to give the *p*-nitrophenol **105**, which was then acetylated and reduced to **106**. This compound was diazotized, reduced with hypophosphorous acid and hydrolyzed to give 2,6-dinonylphenol **107** in 23% yield.

$$(CH_2)_7-CH=CH_2$$

$$O_2N-\!\!\!\!\diagdown\!\!\!\!\diagup\!\!\!-OH$$

$$(CH_2)_7-CH=CH_2$$

105

$$(CH_2)_8-CH_3$$

$$R'-\!\!\!\!\diagdown\!\!\!\!\diagup\!\!\!-OR$$

$$(CH_2)_8-CH_3$$

106: R = Ac, R′ = NH₂
107: R = R′ = H

A further attempt by Lüttringhaus and Vollrath toward the synthesis of catenanes was made on a different system.[82, 83]

According to Ruzicka and Giacomello[105] the spatial arrangement of higher cycloalkanes approximates with increasing number of ring members the shape of a linear double thread molecule. As was first shown by Salomon using kinetic measurements[105a] and later by Lüttringhaus et al.[105b] by preparatory methods, macrocycles may exist in solution as stretched or doubled up entities depending on the solvent. To favorably influence an intraannular conformation between a ring and a chain attached to it, Lüttringhaus and Vollrath introduced rigid segments into the macrocycle. An opening of the ring was thereby achieved.

To obtain such a macrocycle, the resorcinol derivative **108** was reacted with hydroquinone to **109**, which in turn was cyclized to give the macrocycle **110** in 60% yield. The reaction was carried out by the high-dilution method in isoamyl alcohol and in the presence of potassium carbonate. A simpler procedure, in which hydroquinone and the dibromide **108** are reacted in a 1:1 molar ratio, gives a yield of 14–18%. It is interesting to observe that after crystallization from ethanol, solvent molecules are so firmly attached to this compound that they are released only after heating in a vacuum at 150°. The lithium compound obtained by reacting **110** with phenyllithium in ether, gave the carboxylic acid **111** in 43% yield when reacted with carbon dioxide. A similar reaction with ethyl formate gave the aldehyde **112** in 47% yield while *N*-methyl formanilide and ethyl-*o*-formate did not react. The latter two compounds probably did not react due to steric hindrance. On the other hand it was possible, following a procedure given by Wittig *et al.*,[106] to react 2-lithioresorcinol dibutyl ether with *N*-methyl formanilide to obtain the corresponding aldehyde in 50% yield.

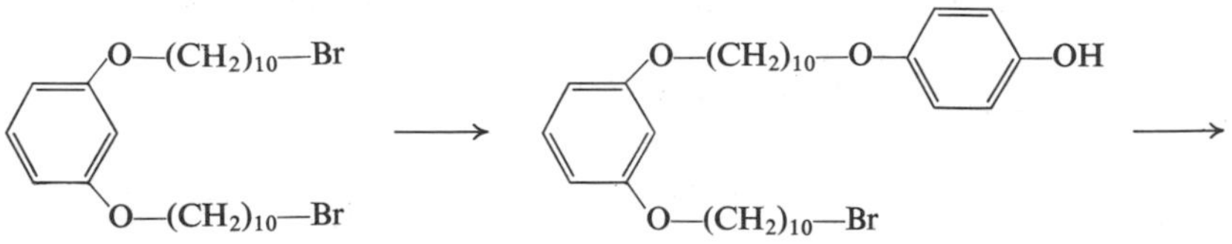

108   109

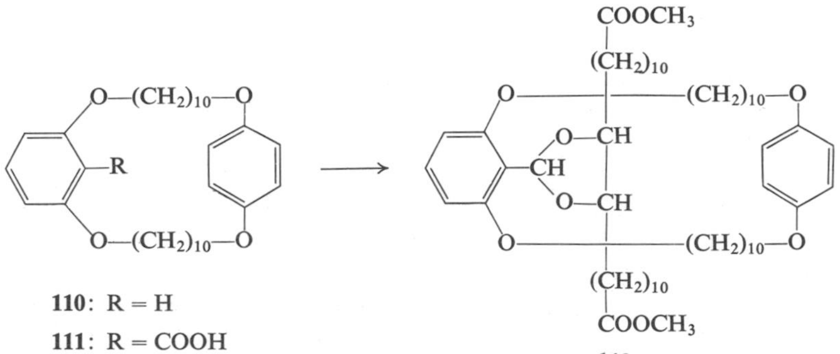

110: R = H
111: R = COOH
112: R = CHO

113

In preliminary experiments it could be shown that 2,6-dimethoxy-benzaldehyde could be ketalized in high yield with *meso*-butane-2,3-diol and *meso*-octane-4,5-diol. The ketalization of aldehyde **112** with dimethyl-*meso*-11,12-dihydroxytetracosanoate[70, 81] according to Salmi[93] gave the ketal **113** in 40% yield.* The formation of the intraannular ketal **113** was hoped to be favored on the one side by the conformation of the macrocycle **112** where the carbonyl group is probably oriented toward the inside of the

* The pictural representation in this and the following formula of similar type are arbitrary and are not meant to specify the actual conformation.

ring, and on the other side by the opening effect on the ring by the two rigid aromatic nuclei.

On cyclizing the ketal **113** with finely divided sodium in xylene at 110°,[11-13] no reaction took place. Most of the unchanged starting material could be recovered. This result was surprising, since in two previous cases ketals have been cyclized without difficulty.[81, 103] The failure to react may be attributed to an exclusive intraannular conformation of **113**, a condition which with the given chain length would strongly hinder the approach of the chain ends to each other. A further cause could be the concentration of oxygen atoms (ether and acetal functions) in the resorcinol part of the molecule. This could bring about a relatively strong attachment of the molecule to the sodium surface and thus prevent solvation. To elucidate these points the ketal **114**, similarly synthesized, was subjected to an acyloin condensation. Since no reaction took place in this case either, one may attribute the failure of the reaction to the second cause given above.

114

Since the acyloin reaction failed with compound **113**, it was attempted to cyclize an analogous dinitrile by the method of Ziegler. In order to prevent the failure of the reaction due to an insufficient chain length, the dibromide **115**[70, 81] was converted to the dimercaptan **116**. This compound was then alkylated with 6-bromocapronitrile to the dinitrile **117**.

In a preliminary reaction this compound was reacted with cyclohexanone to give ketal **118**, which in turn was cyclized after Ziegler. The enaminonitrile **119**, obtained in 14% yield, was characterized as the reaction product **121** with hydroxylamine. The hydrolysis of compound **119** gave in addition to cyclohexanone, the cyano ketone **120**. The reaction of **119** with phenylhydrazine and p-nitrophenylhydrazine in acid solution led to compounds **122** and **123**, respectively.

The reaction of the macrocyclic aldehyde **112** with the dihydroxydinitrile **117** gave the acetal **124** in 61% yield. The viscous oil thus obtained was

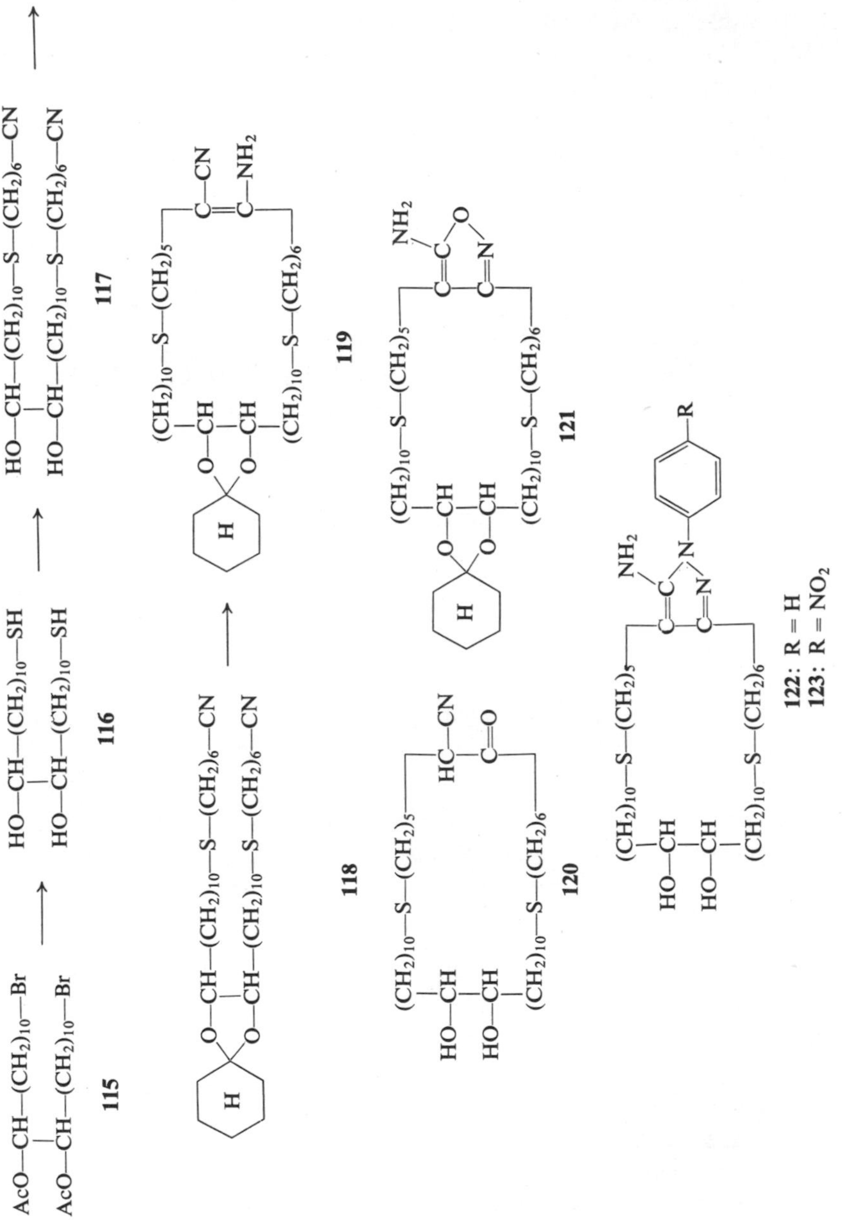

$$AcO-CH-(CH_2)_{10}-Br$$
$$AcO-CH-(CH_2)_{10}-Br$$
115

$$\longrightarrow$$

$$HO-CH-(CH_2)_{10}-SH$$
$$HO-CH-(CH_2)_{10}-SH$$
116

$$\longrightarrow$$

$$HO-CH-(CH_2)_{10}-S-(CH_2)_6-CN$$
$$HO-CH-(CH_2)_{10}-S-(CH_2)_6-CN$$
117

118

119

120

121

122: R = H
123: R = NO$_2$

cyclized according to Ziegler and the reaction product then extracted with petroleum ether. The extracts which contained the enaminonitrile **125**, were hydrolyzed under acidic catalysis. Besides a number of unidentified substances only the macrocyclic aldehyde **112** and the cyanoketone **120** were isolated. No catenane could be detected.

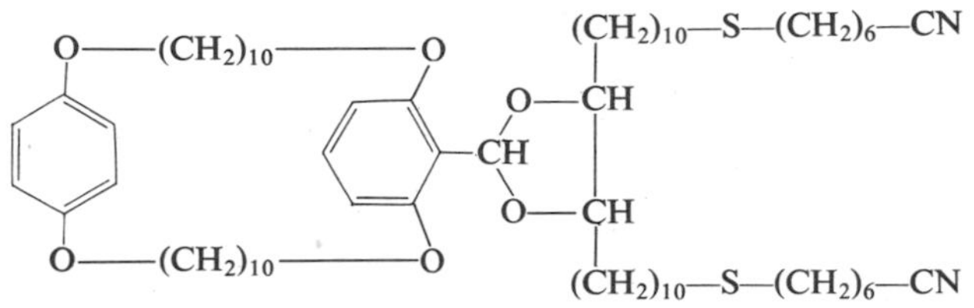

124

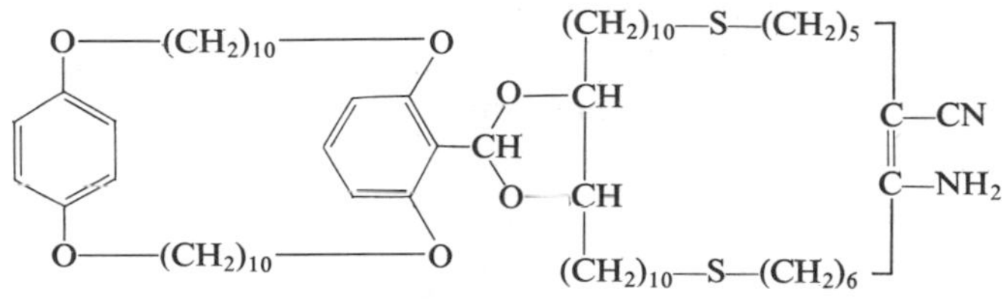

125

Based on Schill's procedure,[15, 16, 107] Lüttringhaus and Isele[20, 21] were able to synthesize a catenane following statistical methods.

Starting from the macrocyclic veratrole ketone **126**[107] the dimethoxy-*m*-cyclophane **127** was obtained by reduction according to Huang-Min-lon[108] and remethylation. The nitration in the 4-position with cupric nitrate in acetanhydride according to Menke,[109] a method which usually gives almost quantitative yields with benzodioxoles,[91] yielded in this case about 70% of compound **128**. A second reaction product was obtained in about 20% yield which, based on its NMR spectrum, probably had the structure **130** or **131**. When this second reaction product was catalytically reduced with Raney nickel, the same product **129** was obtained as when compound **128** was catalytically reduced. The identity of the two compounds was shown by the identical melting points of the *N*-acetyl derivatives.

To verify the 4-position of the nitro group in the nitropyrocatechol dimethyl ether **128**, the nitrocatechol ketal **132**[16] was hydrolyzed to the

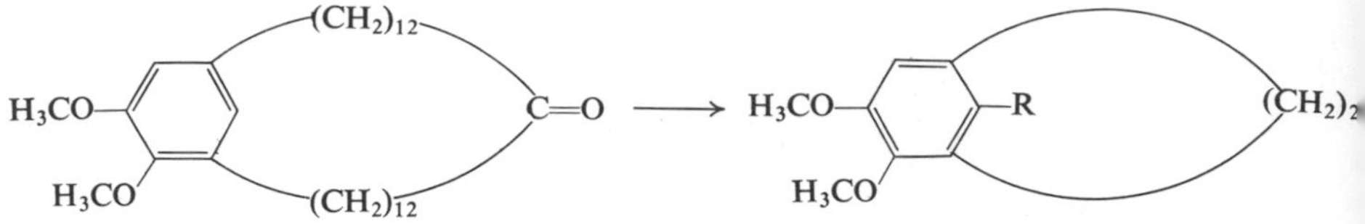

126

127: R = H
128: R = NO$_2$
129: R = NH$_2$

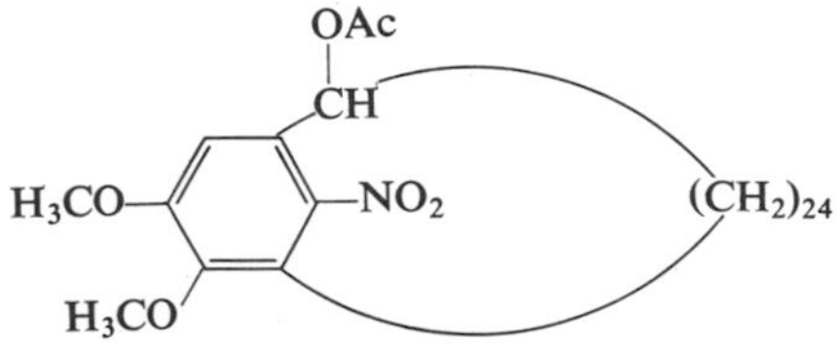

130

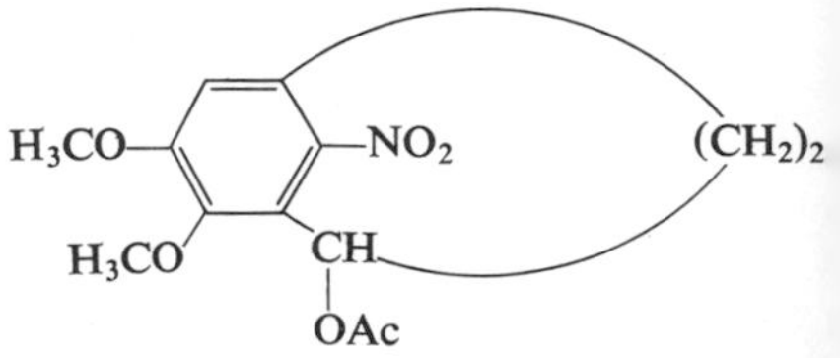

131

nitrocatechol **133** and methylated. The compound thereby obtained was identical with the nitro compound **128**. The position of the nitro group was now evident, since Schill[91] had shown that in 4,6-dialkylbenzodioxoles nitration takes place in the 5-position.

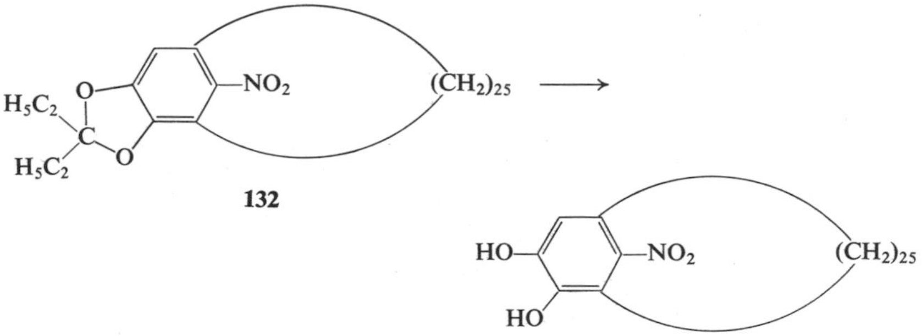

132

133

In the chosen synthetic path it was planned to alkylate the amine **129** with two alkyl chains containing cyclizable end groups. In the catenanes synthesized[16] the second ring contains 26 members, so that the alkylation with 11-membered chains seemed almost insufficient, but just adequate. As in all work with long-chain compounds the availability of the starting material had some influence on the choice. The cyclization was intended to consist of a dinitrile ring closure according to Ziegler, since the acyloin cyclization according to Hansley, Prelog, and Stoll is unsuccessful with catechol derivatives when carried out in the usual manner.[107] Since the usual methods for dialkylation of aniline derivatives of the type **129** with long-chain alkyl halides were not successful,[107] the compound was first acylated with 11-chloroundecanoyl chloride to give **134**. The reduction with diisobutyl aluminum hydride[110] did not take place, even though it usually is a useful reagent for the reduction of mono-substituted amides containing chlorine bound to an aliphatic chain.[15,111] With LiAlH$_4$ the predominant reaction consisted in the reductive elimination of chlorine.

After this unsuccessful reaction sequence, the amine **129** was acylated with 11-(p-methoxyphenoxy)undecanoyl chloride to the amide **135**. The p-methoxyphenoxy group, first used by Ziegler and Weber,[112] already had been found useful in an earlier synthesis of a diansa compound.[113] The alkylation of amide **135** with 11-(p-methoxyphenoxy)undecyl bromide was successfully carried out in dimethylformamide using sodium hydride as a base.[114-116] By the reduction of **136** with LiAlH$_4$ in boiling tetrahydrofuran, amine **137** was obtained. Reaction with hydrobromic acid, followed by methylation, gave the dibromide **138**. This compound could be converted with potassium cyanide to dinitrile **139** (respectively **140**).

Dinitrile **139** (respectively **140**) appeared especially promising for a statistical synthesis of catenanes for two reasons: First, it is possible to cleave the aryl–nitrogen bond in 4-aminocatechol derivatives,[15,16] and second, due to the *ortho* position of the polymethylene bridge, the two N,N-dialkyl residues should, at least in the neighborhood of the amino group, be oriented in a vertical position relative to the plane of the benzene ring. If the polymethylene bridge takes on a conformation in which it is approximately in the plane of the aromatic ring as implied in **139**, the condition is satisfied for an intraannular ring closure, i.e., the formation of a precatenane. It cannot be predicted, however, how great the tendency is for the polymethylene bridge to leave the plane of the benzene ring as implied in **140**.

By using the method of Ziegler together with the high-dilution technique, the dinitrile **139** (respectively **140**) was cyclized in ether with sodium N-methyl anilide. After the hydrolysis of the obtained enaminonitrile with sulfuric acid and remethylation, the reaction product was analyzed with column and thin-layer chromatography. Only the main part of the reaction product, which was equivalent to 52% yield, showed a carbonyl band in the IR spectrum. Proof that this compound was the extraannular ketone **141**, was obtained by cleaving the aryl–nitrogen bond according to the method of Schill.[15,16] The only cleavage products which could be identified, were the macrocyclic hydroxy-p-benzoquinone **142**[16] and the macrocyclic

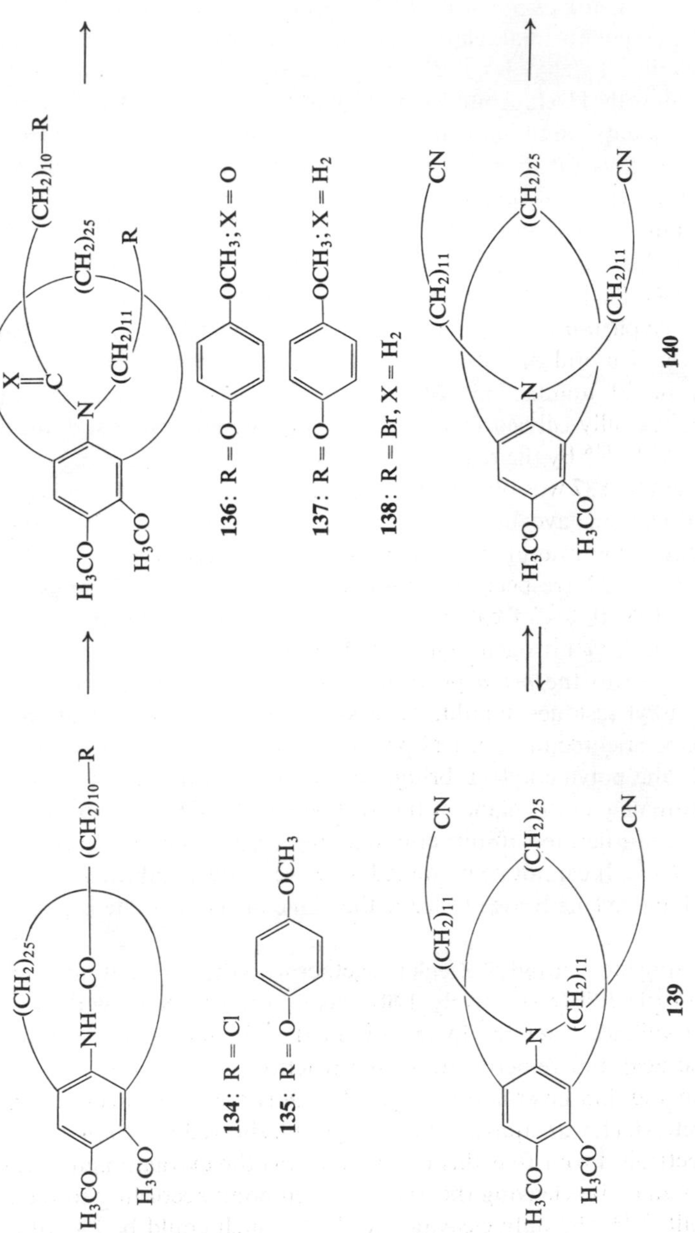

136: R = O—⟨ ⟩—OCH₃; X = O

137: R = O—⟨ ⟩—OCH₃; X = H₂

138: R = Br, X = H₂

140

134: R = Cl

135: R = O—⟨ ⟩—OCH₃

139

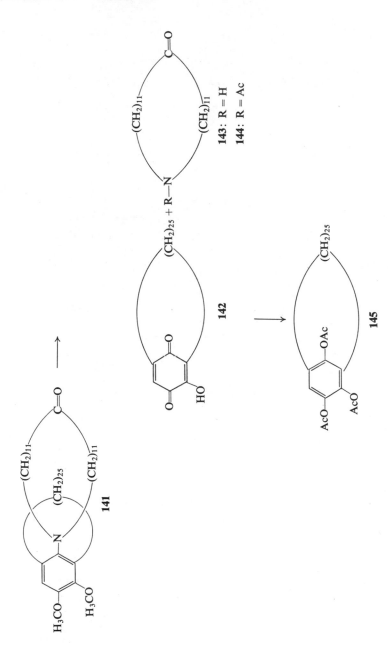

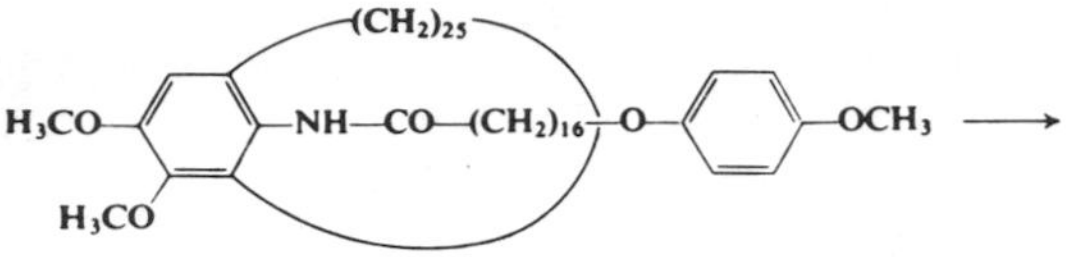

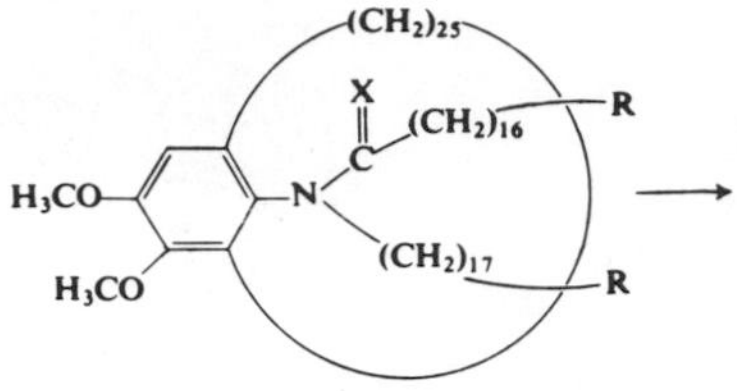

146

147: R = O—⟨benzene⟩—OCH₃; X = O

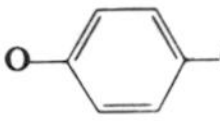

147: R = O—⟨⟩—OCH₃; X = O

148: R = O—⟨⟩—OCH₃; X = H₂

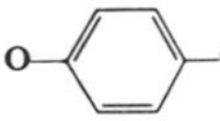

149: R = Br; X = H₂

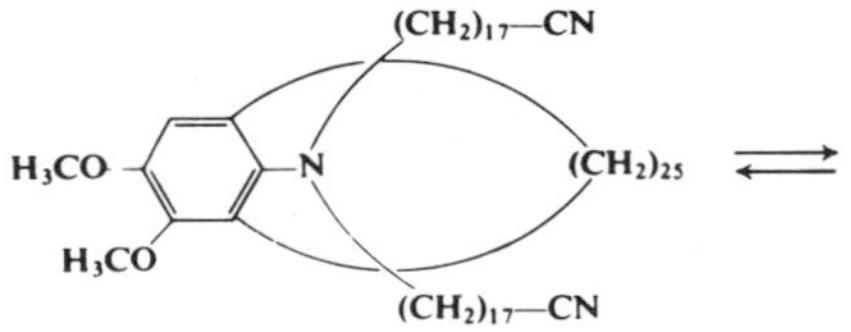

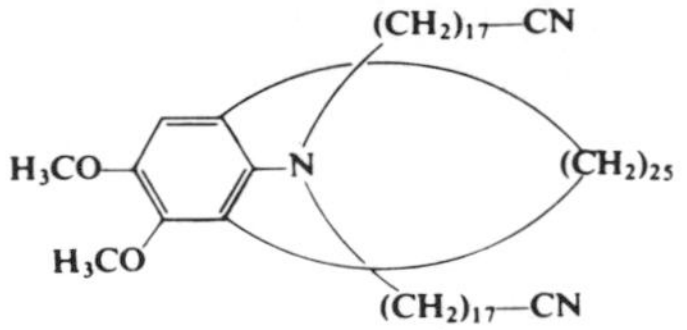

150 ⇌ 151

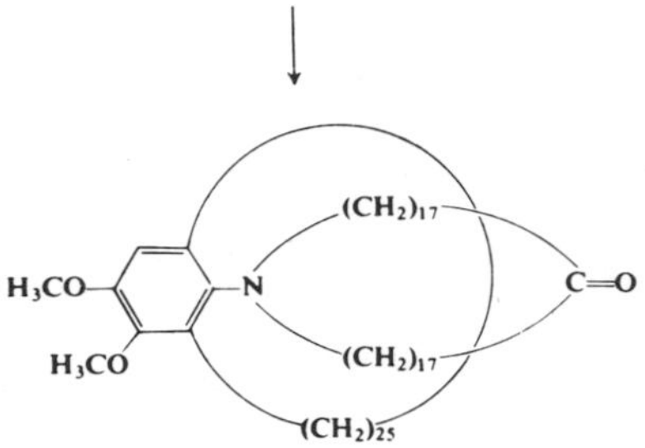

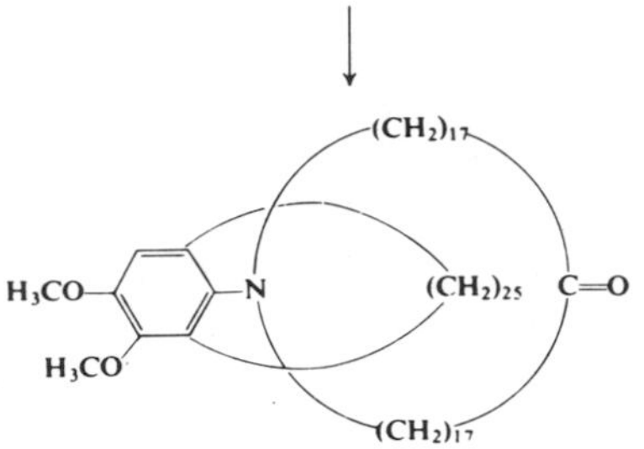

152

153

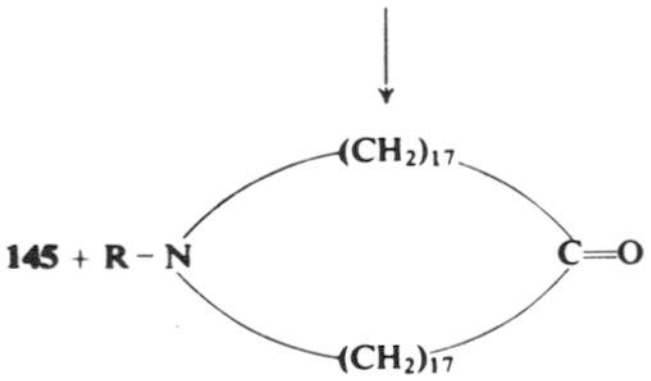

145 + R—N

154: R = H
155: R = COCH₃

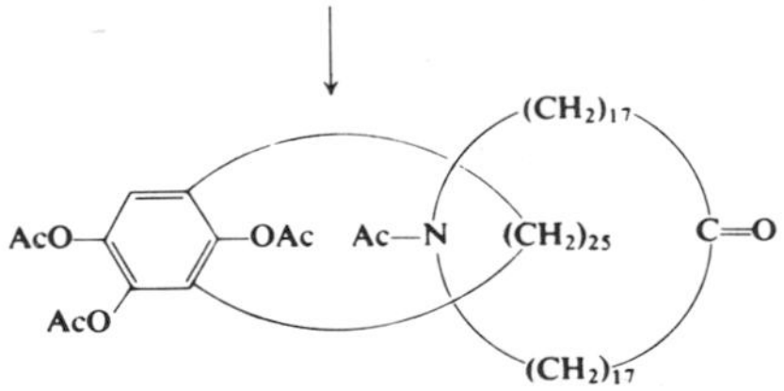

156

azaketone **143**. Because of the unstability of the quinone **142** when subjected to chromatography on silica gel, a portion of the reaction product was reductively acetylated. Besides the triacetate **145**[16] and the *N*-acetyl compound **144**, no catenane could be found after chromatography in this case either.

On the hypothesis that an insufficient chain length of the *N,N*-alkyl substrates in compound **139** was responsible for the exclusive extraannular cyclization, an analogous reaction sequence was carried out starting from the macrocyclic aniline derivative **129**, 17-(*p*-methoxyphenoxy)heptadecanoyl chloride and 17-(*p*-methoxyphenoxy)heptadecyl bromide. As starting material for the latter two compounds 17-bromoheptadecanoic acid was used. With the method found by Hauser and co-workers[117] and extended by Hünig *et al.*,[118] the latter compound was easily obtained from morpholinocyclohexene and 11-acetoxyundecanoyl chloride.[107] By the way of compounds **146–149** the dinitrile **150** was obtained. The latter compound possibly is in equilibrium with its intraannular conformer **151**. The cyclization of the dinitrile gave the isomers **152** and **153**, which could be separated chromatographically on silica gel. They had identical IR spectra. The NMR spectra are similar, showing significant differences, however, in the relative intensities of the split signals due to the polymethylene ring protons of **152** and **153**. The mass spectra of the two isomers are also almost identical. In both compounds a conspicuous peak due to its high intensity appears at m/e 1003. It is the molecular ion peak as well as the base peak.

The total yield of the cyclization was 57%,[20] from which 92–95% was compound **152** and 5–8% precatenane **153**. Renewed verification led to values of 83–85% and 15–17%.[21]

In a further reaction sequence, dibromide **149** was synthesized via compounds **158–162**, starting from the macrocyclic cyclohexanone ketal **157**. Ketal hydrolysis and ether cleavage of **162** gave, after methylation, the desired compound **149**. In the cyclization of the dinitrile thus synthesized, 68% of the extraannular ketone **152** and 32% of the intraannular ketone **153** were obtained. It still remains to be determined whether the shifts in the yields have a specific reason or just were accidental.[21]

The cleavage of the *N*-phenyl bond according to Schill[15, 16] showed that compound **153** was a precatenane and **152** was the extraannular product, respectively. After removal of the methoxyl groups in **152** with hydrobromic acid in propionic acid, and dehydrogenation to the corresponding *o*-quinone, the hydroxybenzoquinone **142** and the macrocycle **154** were obtained by acid hydrolysis. Reductive acetylation of hydroxyquinone **142** led to the triacetate **145**, and acetylation of **154** to the compound **155**. An

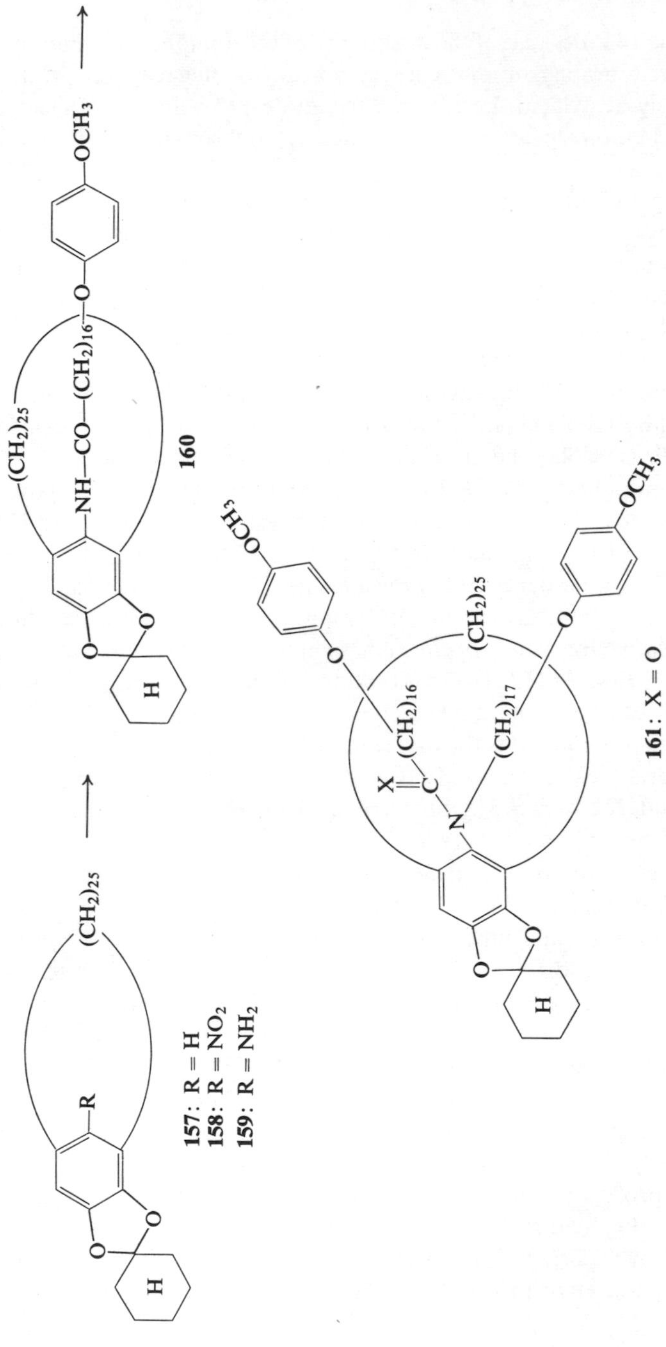

157: R = H
158: R = NO_2
159: R = NH_2

160

161: X = O
162: X = H_2

analogous reaction sequence carried out with compound **153** led, after reductive acetylation, to the catenane **156**. This catenane was obtained as a colorless oil. The mass spectrum of the catenane is completely analogous to the mass spectrum of the catenane obtained by Schill through the directed synthesis method.[19] The mass spectrum is therefore a valuable verification of the proposed structure.

8.2.1. *Investigation of the Equilibrium between Extra- and Intraannular Conformations*

The formation of the two stereoisomers **152** and **153** posed the question of whether the relative amounts of compounds **150** and **151** are fixed during the alkylation of the amide **146** or whether an equilibrium is established between them. A first attempt in answering this question[21] consisted in heating the mixture of dinitriles **150** and **151** in dioxane at 100° for 500 hours before the cyclization. This was done in order to obtain the equilibrium mixture which possibly had not been established before. Dioxane was chosen, because in it paraffin chains tend to exist in a doubled up conformation.[105b] The relative amounts of **152** and **153** did not change during this procedure, however. Although the mixture of dinitriles **150** and **151** gave only one spot in the thin-layer chromatogram, it was observed that amides **147** and **161** always showed two spots, the first one always having a tail. If the chromatography was carried out at 0°, no tail formation was observed. To determine whether this behavior was due to conformeric amides, compound **163** was synthesized. It was found that this compound behaved similarly when analyzed by thin-layer chromatography.

$$(CH_2)_{17}—CH_3$$

$$H_3CO—\qquad—N\qquad (CH_2)_{25}$$

$$C\!=\!O$$

$$H_3CO$$

$$(CH_2)_{16}—CH_3$$

163

From photometric measurements of the equilibrium distribution of the two isomers of compound **163** on thin-layer chromatograms at −32°, −11°, 0°, and +15° a free enthalpy of activation $\Delta G^* = 20.3$ kcal/mole $\pm 3\%$ at

$60°$ was calculated. These values are of the same order of magnitude as those found with NMR measurements by Kessler and Rieker[119] for the *cis–trans* isomerization of some *ortho*-substituted acetanilides. Since only with the amides has there been so far a separation possible, one is led to the assumption that the phenomena studied are not due to conformational isomers but rather to *cis–trans* isomerism of substituted acetanilides.

A different approach toward settling the question of an intraannular–extraannular conformational equilibrium, is to study the optical properties of the system.[21] Amides of the type **136, 147, 161, 163** can occur as optical antipodes in the extra-, as well as intraannular conformation.

In the extraannular conformation the interconversion between antipodes **166** and **167** is only hindered by the substituents in the *ortho* position and should therefore take place readily. In the intraannular conformation the interconversion between **164** and **165** is only possible through an extra-annular conformation and should therefore take place with more difficulty.

To locate optical activity, amide **168** was synthesized and chromato-graphed on cellulose-2,5-acetate.[120] Even at $0°$ no optical activity could be detected. After the failure of this approach, it was attempted to solve the problem via diastereomers. For this purpose acetal **169** was prepared from 3,5-pentacosamethylene pyrocatechole[107] and propionaldehyde. Here again, no optical activity was observed after chromatography on cellulose-2,5-acetate.

Next, the diastereomeric ketals **170** and **171** were synthesized from 3,5-

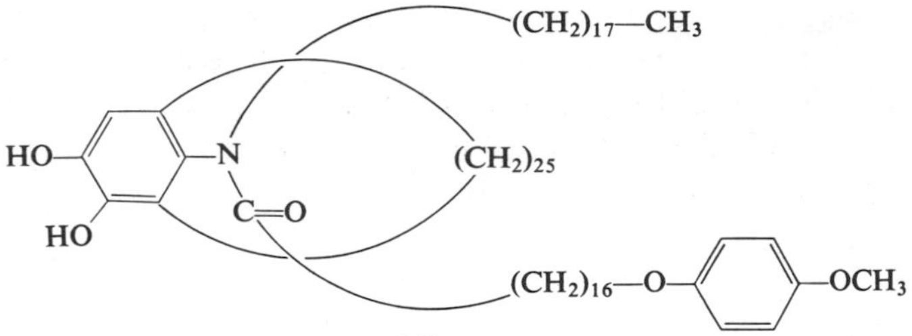

168

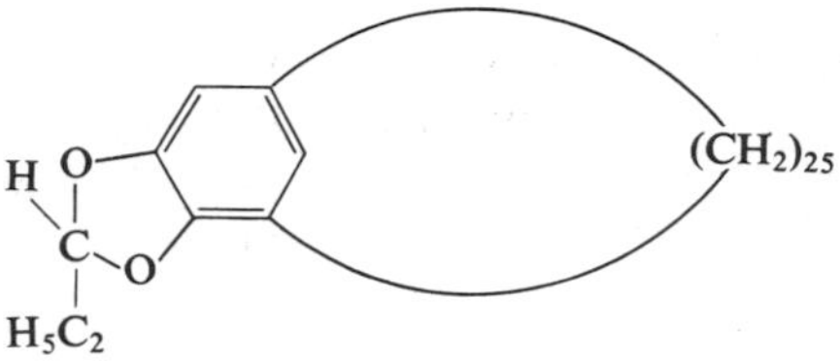

169

pentacosamethylene pyrocatechol and L-menthone, and separated on silica gel. One of the diastereomeric ketals was converted, analogous to the synthesis of **161**, to the amide **172**. A thin-layer chromatogram of this compound on silica gel again showed two isomers. The first measurements in this

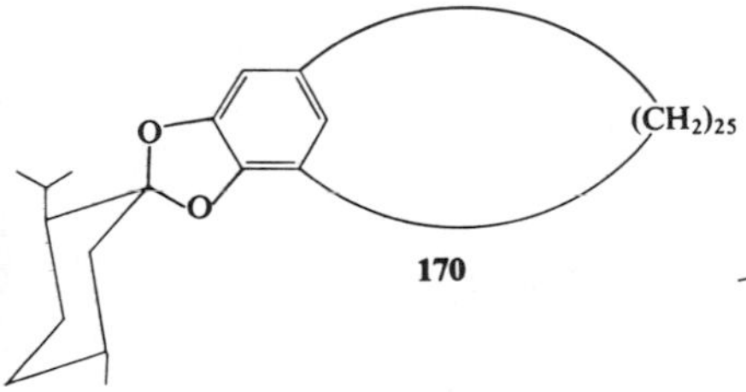

170

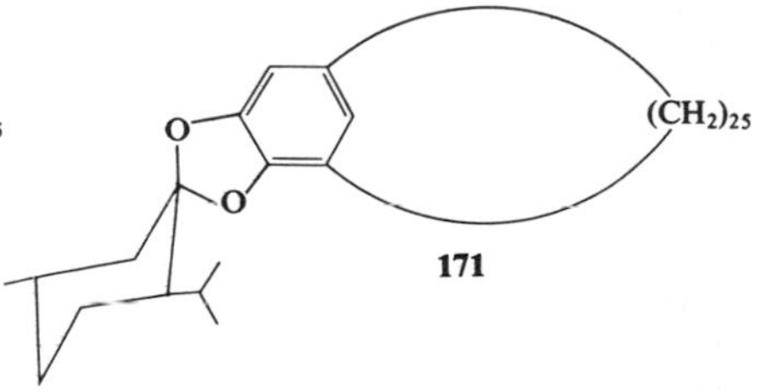

171

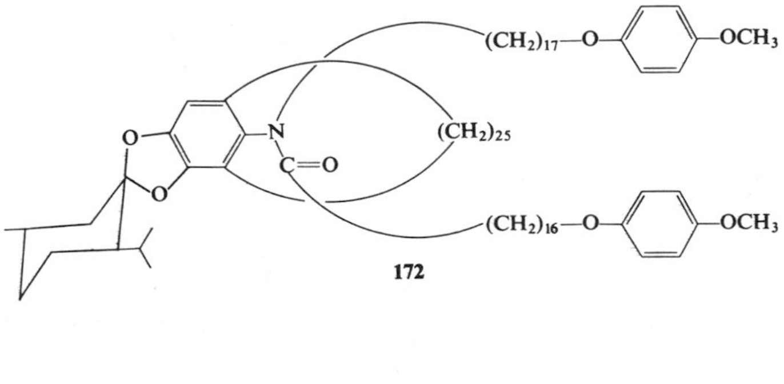

172

experiment, which is not yet concluded, showed no change of rotary power with time. Even in the case of a positive result it would not be certain, however, whether a change in rotary power is due to *cis–trans* isomerism of the amide or whether it is caused by conformational changes, corresponding to a racemization of **164** or **166**.

8.3.   Scheme 3

Lüttringhaus and Winterhalter intended to cyclize sulfides of type **173** ($n > 20$, $m > 20$), to obtain sulfonium salts **174** and **175**. By an adequate substitution of the macrocycle, a cleavage of the sulfonium salts was to be accomplished. Hereby two macrocycles would be obtained.[121]

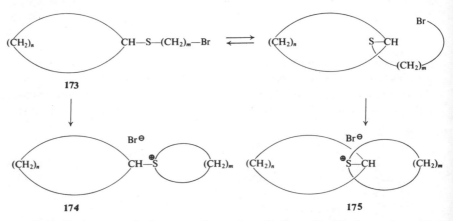

In preliminary work, however, it was found, that no sulfonium salt could be obtained from compound **176**. The same behavior was observed in a similar reaction in which it was attempted to obtain a cyclic ammonium salt from the tertiary amine **177**.

$$Br—(CH_2)_{21}—S—(CH_2)_2—OH \qquad Br—(CH_2)_{21}—N(CH_3)_2$$
$$\textbf{176} \qquad\qquad\qquad \textbf{177}$$

8.4.   Scheme 4

In the years 1956 to 1959, Lüttringhaus and Schill attempted to synthesize catenanes by joining two chains, each containing functional end groups.[70, 81] Cyclization of the end groups and cleavage of the original bonds between the chains was to result in a catenane.

For this purpose 11-bromoundecanoyl chloride was converted by keten dimerization[121a] and hydrolysis to ketodibromide **178**. This compound was converted to dinitrile **179**, hydrolyzed to the diacid **180** and esterified to the ketodiester **181**. The ketodibromide could also be obtained by peroxide catalyzed hydrogen bromide addition to heneicosa-1,20-dien-11-one or by esterification of 1,21-dihydroxyheneicosan-11-one. The latter two compounds can be easily obtained from the keten dimerization of the corresponding acid chlorides.

$$X—(CH_2)_{10}—CO—(CH_2)_{10}—X$$

178: X = Br

179: X = CN

180: X = COOH

181: X = COOCH$_3$

In preliminary work for the ketalization, it was shown that *meso*-11,12-dihydroxydocosa-1,21-diene[92] and heneicosa-1,20-diene-11-one react to give ketal **182**, a mobile oil at room temperature.

$$CH_2=CH—(CH_2)_8—CH—CH—(CH_2)_8—CH=CH_2$$
$$CH_2=CH—(CH_2)_8 —C—(CH_2)_8—CH=CH_2$$
182

The two bifunctional ketones **179** and **181** were ketalized with the bifunctional dihydroxy compounds **85** and **86**, according to Salmi,[93] to give compounds **183** and **184**. A ketal linkage was chosen in order to ensure a perpendicular linkage between the two chains, since it was hoped that this geometry would favorably influence the intraannular cyclization. Furthermore, a ketal provides a temporary linkage which can easily be cleaved again, and which does not interfere in the cyclization methods of Ziegler (see Section 8.2) or the acyloin reaction.[103]

From the cyclization of tetranitrile **183** the monomeric enaminonitriles **185** and **186**, as well as the intraannular analog of **185**, could be expected. To the only distinct reaction product which could be isolated, structure **186** was assigned on the basis of the subsequently mentioned experimental results. Reaction of **186** with hydroxylamine led to the bisisoxazolamine **187**, while acid hydrolysis gave **188**. Compounds **186–188** probably are present as a mixture of structural isomers due to the variability of the position of the nitrile groups. Reaction of **188** with diazomethane in the presence of BF$_3$,[122] followed by hydrolysis with dilute sulfuric acid, resulted in the triketone **189**.

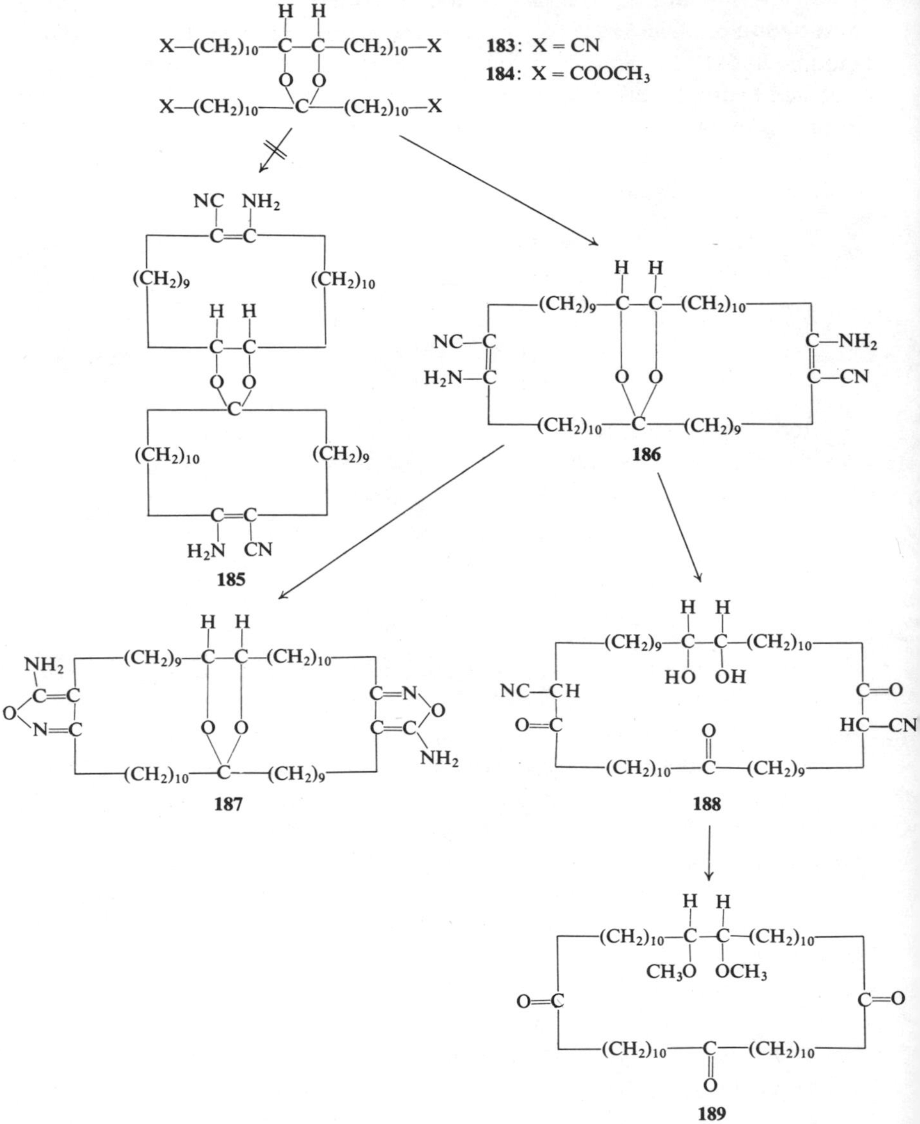

183: X = CN
184: X = COOCH₃

185

186

187

188

189

This compound could not be isolated in pure form. On the basis of the chemical reactions carried out, it could not be decided whether the crystalline cyclization product from tetranitrile **183** was compound **186** or the intraannular analog of compound **185**. The relatively high yield of 25–30% obtained, however, makes the assumption plausible that the reaction product actually was **186** and not a precatenane.

8.5. Scheme 5

This scheme for synthesizing catenanes already contains some features of a directed synthesis. The two macrocycles, which will give a catenane or two separate rings after cleaving the bonds linking them are already joined at two-positions. The directed synthesis differs from Scheme 5 mainly in that additional requirements are fulfilled to compel an intraannular cyclization.

The possibilities of synthesizing catenanes after Scheme 5 were first described and experimentally attempted by Schill[15] and later extended by Lüttringhaus and Linke.[111, 123] Further research was begun by Lüttringhaus and Vollrath.[84]

Schill intended to obtain a catenane by the synthesis of a triansa compound of the type **190** ($n > 10$, $m > 20$) followed by the cleavage of the benzyl–nitrogen bonds. One could expect a priori, that considerable amounts of extraannular compound would form during the cyclization of the precursor of **190**.

190

In preliminary work on a method to cleave the double-ansa system, Schill[15] and later Lüttringhaus and Linke,[111] were able to show that the acetolytic cleavage of Mannich bases[124] could be extended to 2,5-bisdialkyl-aminohydroquinone ethers such as **191** and **192**. After refluxing with acetic anhydride, the cleavage products **193** and **195**, as well as **194** and **195**, could be isolated in high yield.

191: $R = CH_3$
192: $R = n\text{-}C_4H_9$

193: $R = CH_3$
194: $R = n\text{-}C_4H_9$

In preliminary work on the cyclization and synthesis of starting materials, it was found that by using the method of Friedman and Shechter[125] dinitriles **196** and **197** could be obtained from the dibromides in high yields. The reduction to the amino compounds **198** and **199** was carried out catalytically with Raney nickel in methanol saturated with ammonia or with LiAlH₄. Acylation with 11-chloroundecanoyl chloride led to the amides **200** and **201**.[15,111,126] These could be reduced to amines **202** and **203** without reductive dehalogenation when diisobutyl aluminum hydride was used. It was found to special advantage to carry out the reduction with diborane in tetrahydrofuran.[127,128]

For all cyclizations, a *N,N*-polymethylene chain of 11 carbon atoms was chosen. This length allows the formation of a diansa compound of type **190**, but due to Pitzer strain and transannular strain, hinders the formation of a tricyclic compound of the type **208**.

Diamine **203** was cyclized in isoamyl alcohol in the presence of potassium carbonate and sodium iodide. The reaction, carried out under high-dilution

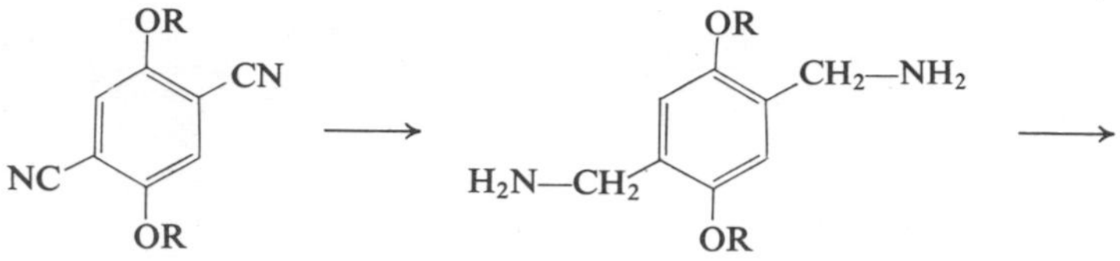

196: R = CH₃
197: R = n-C₄H₉

198: R = CH₃
199: R = n-C₄H₉

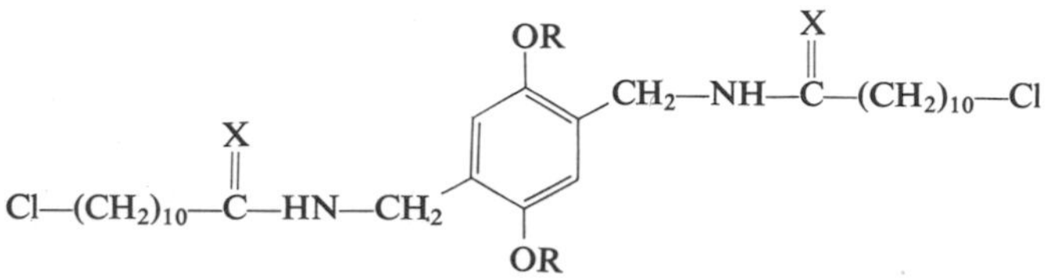

200: R = CH₃; X = O
201: R = n-C₄H₉; X = O
202: R = CH₃; X = H₂
203: R = n-C₄H₉; X = H₂

conditions gave, after separating the polymeric product and acetolysis, the cleavage products **194** and **204** in 70 and 38% yields, respectively.[123]

203 ⟶

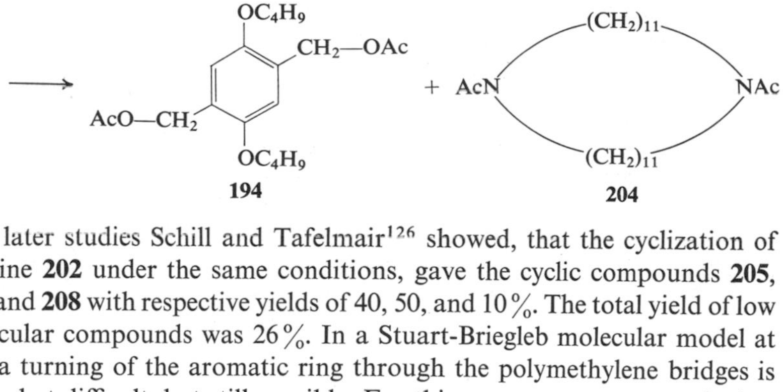

194   204

In later studies Schill and Tafelmair[126] showed, that the cyclization of diamine **202** under the same conditions, gave the cyclic compounds **205**, **206**, and **208** with respective yields of 40, 50, and 10%. The total yield of low molecular compounds was 26%. In a Stuart-Briegleb molecular model at **205**, a turning of the aromatic ring through the polymethylene bridges is somewhat difficult, but still possible. For this reason no attempt to isolate isomers of this compound was made.

Due to the similarity of the chromatographic R_f values, the products of the cyclization were not separated in a preparatory scale. The mixture of the low molecular products from the cyclization gave on acetolysis, diamide **204**, tetraamide **207**, monoamide **209**, and diacetate **193** in the yields of 30, 33, 5.9, and 52%, respectively.

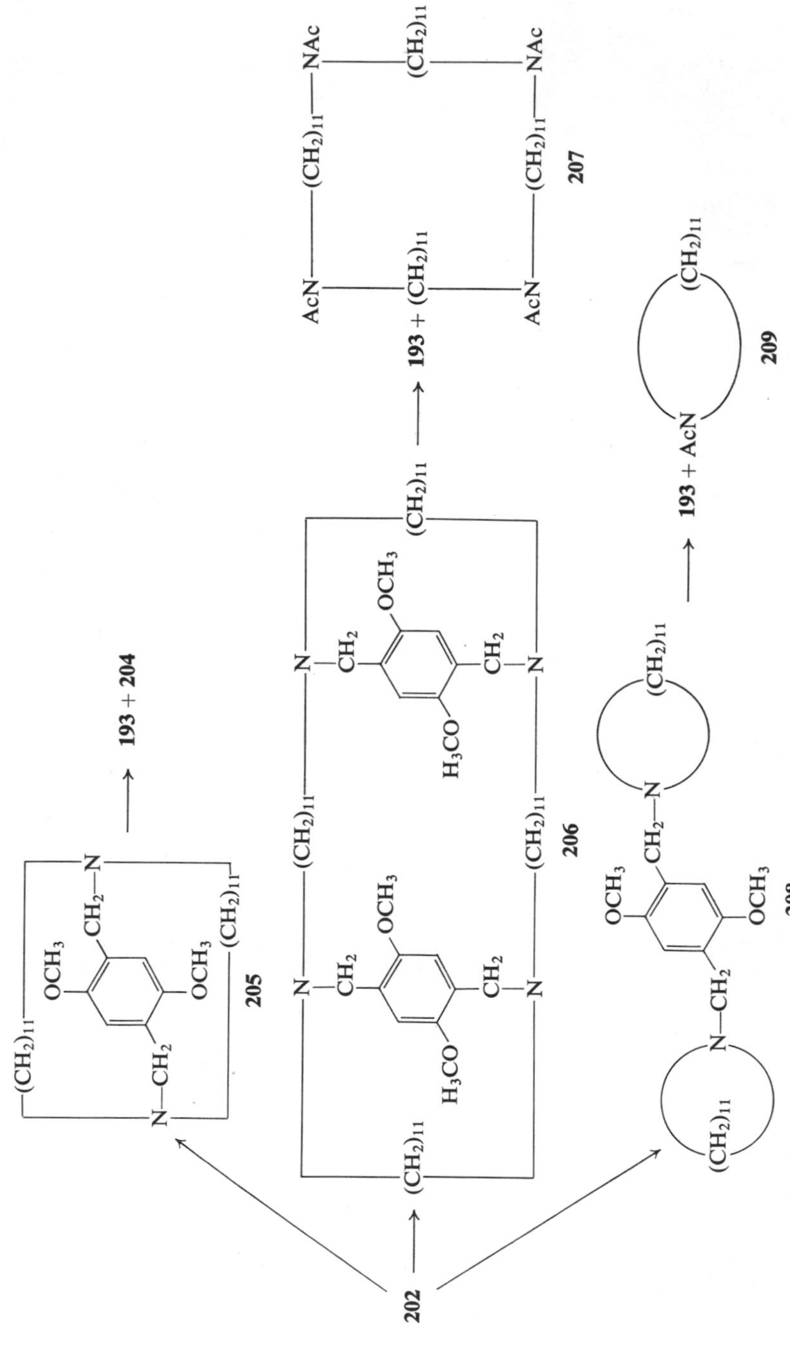

In preliminary work it was shown that cleavage of the cyclization products with cyanogen bromide or hydrogenolysis with palladium on charcoal is not as suitable.[126]

The cyclization products **205**, **206**, and **208** were identified by means of mass spectrometry. Compound **208** was also compared with an authentic sample prepared by the alkylation of cycloundecylamine with 2,5-bis-chloromethyl hydroquinone dimethyl ether. The structure of the macrocyclic tetraamide **207** is evident from the mass spectrum, since macrocyclic amides show typical fragmentation patterns.[19]

Based on the results of the cyclization of diamine **202**, it is probable that the cyclization of **203**[111] led to similar results.

As starting materials for the synthesis of a triansa compound of type **190**, dinitrile **210**,[69] and diester **215**[111] were used. They were easily obtained by

210   **211**   **212**: R = H
                              213: R = Br
                              214: R = CN

215   **216**   **217**: R = H
                              218: R = Br
                              219: R = CN

220: $n = 21$
221: $n = 22$

222: X = O
223: X = H$_2$

alkylating hydroquinone with 11-bromoundecanonitrile or methyl 11-bromoundecanoate. By cyclizing dinitrile **210** according to Ziegler and hydrolyzing the product, ketone **211** was obtained. This compound was converted via Huang-Minlon reduction to the ansa ether **212**.[69] An acyloin condensation of **215** led to the cyclic acyloin **216**, from which **217** was obtained by a Clemmensen reduction.[111] Bromination of the two ansa ethers **212** and **217** gave dibromides **213** and **218**, which were then converted to dinitriles **214** and **219**[15, 111] by the method of Friedman and Shechter.[125]

Catalytic reduction of dinitriles **214** and **219** with Raney nickel in methanol saturated with ammonia led to diamines **220** and **221**. Acylation of **221** with 11-chloroundecanoyl chloride gave diamide **222**, which was reduced with diisobutyl aluminum hydride to diamine **223**.

The cyclization of the macrocyclic diamine **223** in isoamyl alcohol using the high-dilution method in the presence of potassium carbonate and sodium iodide only gave a reaction mixture difficult to purify. No distinct product could be isolated.[123]

After cyclization of diamine **223** and cleavage one could expect to obtain, from the extraannular cyclized portion of the product, the hydroquinone ether **227** and the macroheterocycle **204**. Both cleavage products were independently synthesized in order to better identify these in the chromatographic analysis and to facilitate the differentiation from any catenane formed.

224: R = COOH
225: R = COOCH$_3$
226: R = CH$_2$OH
227: R = CH$_2$OAc
228: R = CH$_2$Cl

In order to synthesize **227**, dinitrile **219** was saponified with potassium hydroxide in diethylene glycol to the diacid **224**, which in turn was esterified to **225**. Reduction with LiAlH$_4$ to **226** followed by acetylation led to the desired compound. A parallel method of obtaining **227** consisted in reacting bischloromethyl compound **228**, obtained by chloromethylation of **217**, with sodium acetate.

The macroheterocycle **204** was obtained by three different methods:

1. From *p*-toluenesulfonamide and 1,11-dibromoundecane in a molar ratio 1:1.
2. From (11-bromoundecyl)-*p*-toluenesulfonamide.
3. From 1,11-bis-*p*-toluenesulfonamidoundecane and 1,11-dibromo-undecane in a molar ratio 1:1.

All cyclizations were carried out using the high-dilution method in the presence of potassium carbonate and sodium iodide. The yields from these three methods were 18, 65, and 60%, respectively. Cleavage of compound **229** with hydrobromic acid or better yet with LiAlH$_4$ in tetrahydrofuran gave compound **230**, which upon acetylation led to compound **204**.

$$2H_3C-\langle C_6H_4\rangle-SO_2NH_2 + 2Br-(CH_2)_{11}-Br$$

$$2H_3C-\langle C_6H_4\rangle-SO_2NH-(CH_2)_{11}-Br \longrightarrow$$

$$H_3C-\langle C_6H_4\rangle-SO_2NH-(CH_2)_{11}-NH-O_2S-\langle C_6H_4\rangle-CH_3$$
$$+$$
$$Br-(CH_2)_{11}-Br$$

macrocycle: $R-N$ ...(CH$_2$)$_{11}$... $N-R$

229: R = Tosyl
230: R = H
204: R = Ac

In another study, Lüttringhaus and Linke[111,123] attempted to synthesize the precatenane **231**. It was to be converted to a catenane by the cleavage of the ketal function and the acetolysis of the benzyl–nitrogen bond.

231

The main steps of the reaction sequence were studied in preliminary reactions. For this purpose the toluhydroquinone dimethyl and dibutyl ether were brominated to **232** and **233**, and reacted with CuCN in dimethylformamide to **234** and **235**. Catalytic reduction in acetic anhydride led to amides **236** and **237**. The attempt to hydrolyze these amides, however, only resulted in resinous products. Reduction of the nitriles with $LiAlH_4$, on the other hand, yielded the amines **238** and **239** in a smooth reaction.

232: R = CH₃; X = Br
233: R = n-C₄H₉; X = Br
234: R = CH₃; X = CN
235: R = n-C₄H₉; X = CN

236: R = CH₃; R′ = COCH₃
237: R = n-C₄H₉; R′ = COCH₃
238: R = CH₃; R′ = H
239: R = n-C₄H₉; R′ = H

In further preliminary studies, the tertiary amines **240** and **241** were cleaved in boiling acetic anhydride to give, in addition to *N,N*-dibutyl-acetamide, compounds **242** and **243** in 75 and 60% yields, respectively.

240: R = CH₃
241: R = n-C₄H₉

242: R = CH₃
243: R = n-C₄H₉

In order to transfer the reaction sequence of compounds **232–238** to a macrocyclic ansa ether, toluhydroquinone was first alkylated with 1,10-dibromodecane to dibromide **244** and then converted to the dinitrile **245**. Cyclization by the Ziegler method and hydrolysis, gave ketone **246** in 44% yield. This compound was also obtained by cyclizing 2,2-bis(10-bromo-decyl)-4,5-dimethyldioxolane with toluhydroquinone in isoamyl alcohol in the presence of potassium carbonate.

Since it was found that bromination of ketone **246** was accompanied by a substitution in the α-position to the carbonyl group, the compound was first converted to the oxime **247**. Bromination to **248** and subsequent hydrolysis gave ketone **249**. Reaction with CuCN in dimethylformamide gave nitrile **250**, which was reduced with $LiAlH_4$ to compound **251**, after having protected the keto group by ketalization with ethylene glycol. Reduction and hydrolysis led to the amine **252**, which could easily be characterized as the chloride or perchlorate. The ketone **252** thus obtained, was ketalized with 1,22-dibromo-11,12-dihydroxydocosane[70] according to Salmi.[93] The resulting ketal **253** was cyclized in isoamyl alcohol in the presence of potassium carbonate. The reaction products obtained were first treated with acid to bring about a ketal cleavage, and then heated in boiling acetic anhydride to cleave the benzyl–nitrogen linkage. Although the reaction product could be separated into a number of fractions when chromatographed on Al_2O_3, no conclusions could be drawn about the structure of the compounds obtained.

O—$(CH_2)_{10}$—X H_3C ... O—$(CH_2)_{10}$—X

244: X = Br
245: X = CN

H_3C ... O—$(CH_2)_{10}$— ... C=X ... R ... O—$(CH_2)_{10}$—

246: R = H; X = O
247: R = H; X = N—OH
248: R = Br; X = N—OH
249: R = Br; X = O
250: R = CN; X = O
251: R = CH_2—NH_2; X = $\begin{smallmatrix} O—CH_2 \\ O—CH_2 \end{smallmatrix}$
252: R = CH_2—NH_2; X = O

H_3C ... O—$(CH_2)_{10}$— ... CH_2—NH_2 ... O—$(CH_2)_{10}$—

Br | $(CH_2)_{10}$ | O—C—H | O—C—H | $(CH_2)_{10}$ | Br

253

Work was also devoted to the synthesis of the molecular subunits resulting from the acetolysis and ketal cleavage of the hypothetical compound **231**. For this purpose, aniline was twice alkylated with methyl 11-bromo-undecanoate, the resulting compound **254** cyclized to the acyloin **255** and subsequently reduced to **256** with $LiAlH_4$. The higher melting dihydroxy compound was converted, using isoamyl nitrite, to the nitroso compound **257**. The cleavage of this compound with alkali[129] or sodium bisulfite[130] has not been successful thus far.

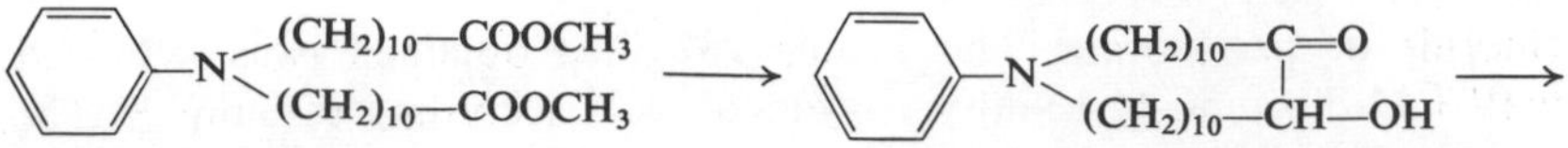

254

255

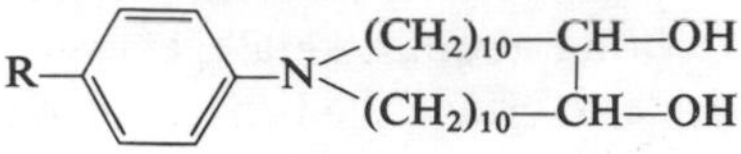

256: R = H
257: R = NO

Lüttringhaus and Vollrath tried to obtain a catenane by the following method.[84] The intraannularly bridged ring system **258**, having 26 links in the ring, was to be twice alkylated with 11-bromoundecanentirile to **259**. A ring closure by the Ziegler method followed by hydrolysis of the resulting enaminonitrile, was to lead to ketone **260** which upon hydrolysis of the amide bonds was expected to give a catenane.

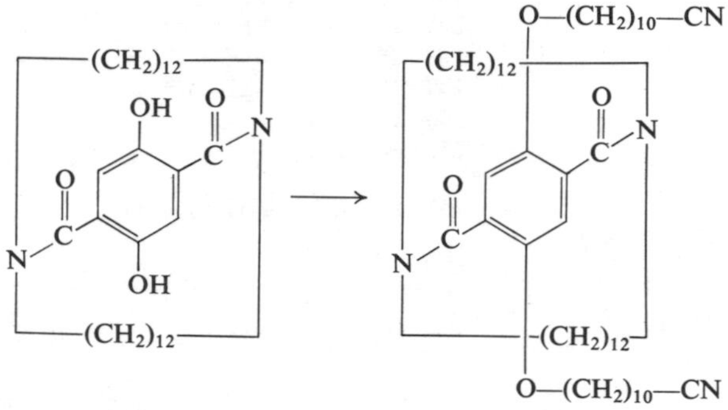

258

259

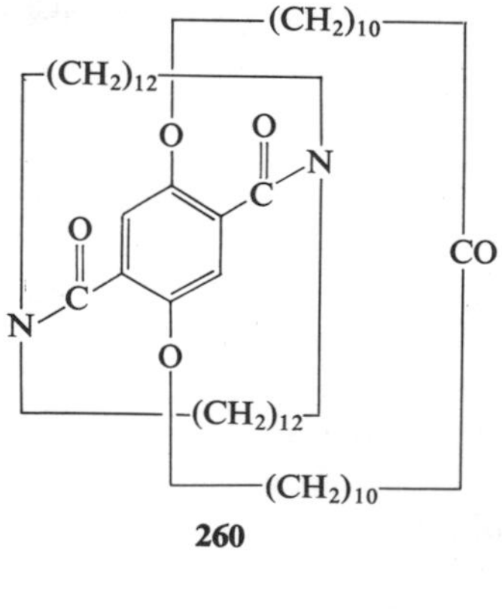

260

The planned reaction sequence was first tested with preliminary reactions. Via ether cleavage with $AlBr_3$, 2,5-dimethoxy-N,N,N',N'-tetrabutyltereph-thalamide **261** was converted to the hydroquinone derivative **262**. Alkylation with 11-bromoundecanenitrile in ethanol, with sodium ethoxide as a base, gave the dialkyl ether **263** in 61 % yield. This compound was cyclized to the enaminonitrile which could be hydrolyzed, without altering the dibutylamide groups, in 70% sulfuric acid to ketone **264**. Since the amide could not be hydrolyzed with other methods either, the compound was reduced with $LiAlH_4$ to compound **265**. Acetolytic cleavage[15] gave, besides the N,N-dibutylacetamide, the triacetate **266**. The latter was identified after hydrolysis as the trihydroxy compound **267**.

261: R = CH_3
262: R = H
263: R = $(CH_2)_{10}$—CN

264: X = O; Y = O
265: X = H_2; Y = H, OH

266: R = Ac
267: R = H

After preparing compound **268** in analogy to a method of Stetter and Marx,[90] it was attempted to synthesize the bridged ring system **269** by reacting the macroheterocycle **268** with 2,5-dimethoxyterephthalic acid chloride under high dilution conditions. It was not possible, however, to isolate the desired compound. Reaction of the macroheterocycle **268** with 2,5-dimethoxy-4-chloroformylbenzoic acid methyl ester led to the mono-amide **270** and the corresponding diamide. After the ester was saponified and converted to the acid chloride, a renewed cyclization did not lead to **269** either. This reaction sequence has not been further investigated.

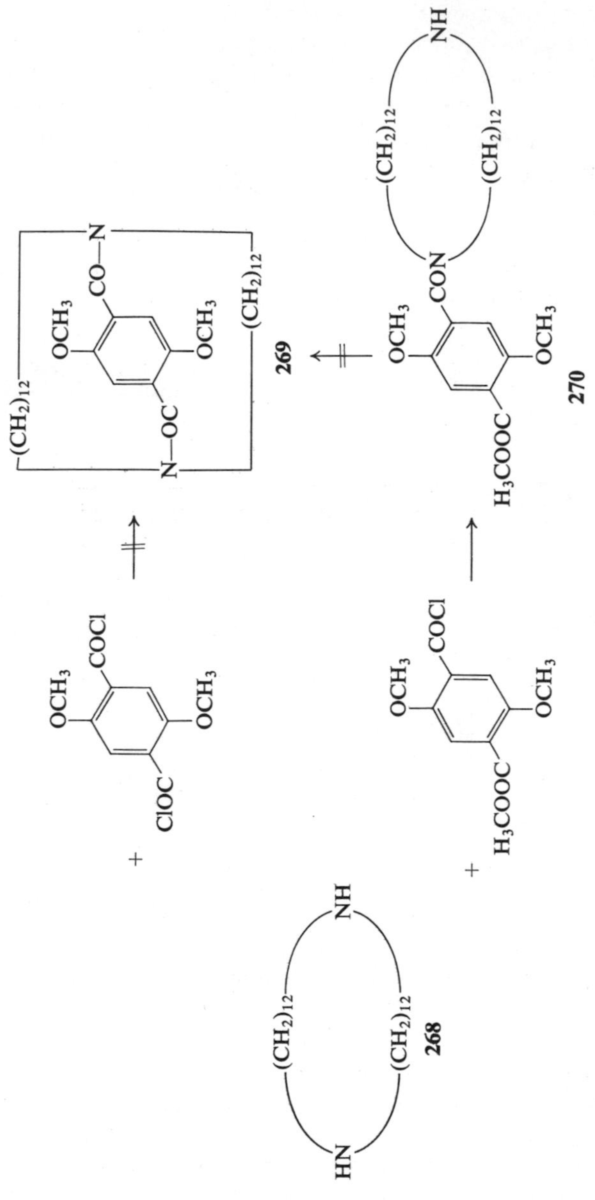

268 269 270

9

Investigations
of the
Directed
Syntheses of
[2]-Catenanes

9.1. Triansa Compounds from 2,5-Diaminohydroquinone Polymethylene Ethers

9.1.1. *Diansa Compounds from 1,4-Dialkoxy-2,5-diaminobenzene Derivatives*

The path for the directed synthesis, as shown in Fig. 2 (Section 7.2), can proceed via a triansa compound. This compound can be considered a precatenane, which upon rupture of the appropriate bonds between the aromatic nucleus and the double-bridge system, becomes a catenane. Thus, the essential problem in a directed catenane synthesis lies in (1) the synthesis of an intraannularly linked precatenane and (2) the cleavage of the bonds between the two ring systems. Since adequate systems had not been investigated, model substances were studied to elucidate the problems. In planning the synthetic paths, consideration was always given to the availability of the starting materials, since at least two ring closures with often moderate yield would be necessary. Furthermore, the stereochemical requirements and the compatability of the system to the planned cleavage and ring closure methods had to be taken into consideration.

Before starting the actual experimental work it had to be decided whether the triansa compound should be obtained by cyclizing **62**, or whether it would be easier to attach two additional bridges to the monoansa compound **65**. To fix a third bridge on a diansa compound seemed more difficult because the bonds between A and B might break prematurely. However, both possibilities for the synthesis of a triansa compound were experimentally tested by Schill,[15, 69] although no attempts were made in this preliminary work to fix the chain perpendicularly to the aromatic nucleus.

A short time after the synthesis of the triansa compound **314** (p. 84) by Schill,[69] the synthesis of the same substance was published by Doornbos and Strating.[131] Both groups worked on model systems before attempting the actual synthesis.

A suitable reaction path for the synthesis of a triansa compound was indicated through the synthesis of the diansa compound **276** by Lüttringhaus and Simon.[113]

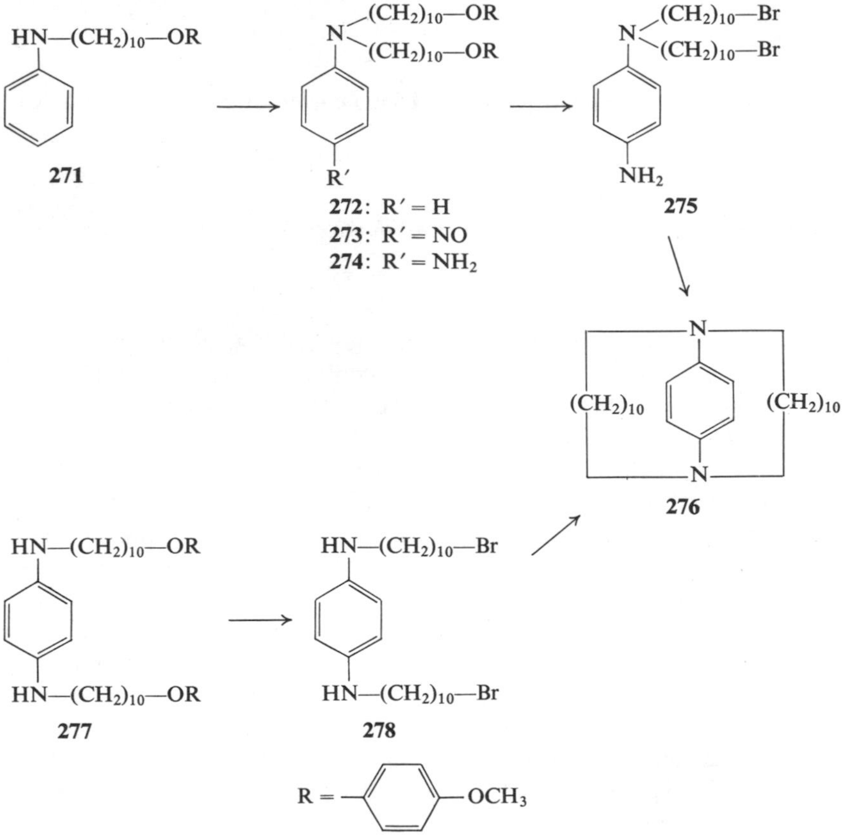

271

272: R′ = H
273: R′ = NO
274: R′ = NH$_2$

275

276

277

278

$$R = \text{—} \bigcirc \text{—OCH}_3$$

In the first method for the synthesis of this compound, aniline was metalated with phenylsodium and then alkylated with (4-methoxy)phenyloxydecylbromide to give **271**. Repetition of these two steps gave the N,N-dialkylaniline derivative **272**. Nitrosation to **273**, followed by reduction, resulted in **274**. Cleavage of the ethers with hydrobromic acid led to **275**, which on cyclization in isoamyl alcohol with potassium carbonate gave diansa compound **276** in 18% yield.

In the second method, the disodium derivative of p-phenylenediamine was reacted with (4-methoxy)phenoxydecylbromide to **277**. Cleavage of the ether led to **278**, which upon cyclization gave **276** in 46% yield.

Doornbos and Strating[38] prepared diamine **277** by alkylation of N,N'-ditosyl-p-phenylenediamine with (4-methoxy)phenoxydecylbromide, followed by detosylation with hydrogen bromide in glacial acetic acid and phenol.[132]

Schill, as well as Doornbos and Strating, attached the third ansa bridge via ether linkages. At first Doornbos and Strating also attempted to obtain the triansa compound from 2,5-dibromo-p-phenylenediamine, but due to difficulties in preparation the reaction sequence could not be concluded.

With the intention of obtaining a 2,5-dialkoxy-p-phenylenediamine diansa compound, Schill started with the model compound 2,5-diethoxyaniline.[15, 69] By acylation with 11-bromoundecanoyl chloride to **279a**, nitration with nitric acid in acetic acid to **280a**, reduction with stannous chloride to amine **281a**, and renewed acylation with 11-bromoundecanoyl chloride, diamide **282a** was finally obtained.

OR
NH—CO—(CH$_2$)$_{10}$—Br
X
OR

$\longrightarrow$

279a, b: X = H
280a, b: X = NO$_2$
281a, b: X = NH$_2$

OR
NH—CO—(CH$_2$)$_{10}$—Br
Br—(CH$_2$)$_{10}$—OC—HN
OR
282a, b

a: R = C$_2$H$_5$
b: R = CH$_2$—CH=CH$_2$

With the intention of introducing by a Claisen allylic ether rearrangement, alkyl substrates in the 3- and 6-positions, a similar reaction sequence was carried out starting with 2,5-diallyloxyaniline (279b–282b). In this manner an attachment of the third ansa bridge would have been possible. These investigations were already linked with the concept of obtaining diansa compounds which would allow cleavage of the bonds between the aromatic nucleus and the double-bridge system (compare Section 9.4.1).

To continue the synthesis, it was planned to reduce the amides 282a, b to the corresponding diamines which then would be cyclized to the diansa compounds. In a preliminary reaction, monoamide 279a could be reduced in 65% yield with $LiAlH_4$ at 0° to amine 283 without reductive elimination of bromine taking place. Amides 282a, b could not be reduced by this method, however.

279a ⟶

OC_2H_5

NH—$(CH_2)_{11}$—Br

OC_2H_5

283

Since at the time no other reduction methods were available, the 2,5-diethoxy-*p*-phenylenediamine diansa compound 287b was synthesized by a different path. By modifying and simplifying the reaction sequence of Lüttringhaus and Simon,[113] it was possible to synthesize the diansa compound 287b.[69] 2,5-Diethoxyaniline was twice alkylated with excess 1,10-dibromodecane. The use of excess dibromide prevented most of the polymer formation. Compound 284b thus obtained, was reacted with nitrous acid to 285b. The position of the nitroso group was not proved, since normally, nitrozation takes place in the *para* position to the dialkylamino group.[113] Amine 286b, obtained by the reduction of the nitroso group with stannous chloride, was cyclized under high-dilution conditions in isoamyl alcohol and in the presence of potassium carbonate. The total yield of the crystalline diansa compound 287b was 17% yield (based on 2,5-diethoxy-aniline).

Starting from 2,5-dimethoxyaniline Doornbos and Strating synthesized in a similar manner (284a–286a) diansa compound 287a in 5.8% total yield.[38]

The (4-methoxy)phenoxy protecting group, used by Lüttringhaus and Simon,[113] is difficult to remove in the presence of other functional groups. The Dutch authors, in their preliminary investigations toward the synthesis of a triansa compound, attempted to by-pass the difficulties encountered with the use of the previously mentioned protecting group in the following manner. Starting with the alkylation of N,N'-ditosyl-1,4-diaminobenzenes **288** and **289** with excess 1,10-dibromodecane compounds **291** and **292** were

284a, b: X = H
285a, b: X = NO
286a, b: X = NH$_2$

287a, b

a: R = CH$_3$
b: R = C$_2$H$_5$

obtained. As a by-product of the alkylation, small amounts of ansa compounds **295** and **296** could be isolated. In earlier work, Stetter and Roos already had synthesized (5-bromopentyl)-N,N'-ditosyl-p-diaminobenzene in a similar manner.[133]

Detosylation of the bissulfonamide **292** to **294** could be accomplished with hydrogen bromide in glacial acetic acid and phenol.[132]

Cyclization of diamine **294** in the presence of potassium carbonate using the high-dilution method in isoamyl alcohol, pentan-1-ol, or dimethylformamide did not lead to the corresponding diansa compound.

In the literature, a number of examples can be found where macrocycles are obtained by cyclizing equimolar amounts of N,N-bisaryl- or alkyl-sulfonamides with α,ω-dihalides.[133-138] Based on these methods, bistosylamides **288** and **289** were cyclized with 1,10-dibromodecane and potassium carbonate in dimethylformamide. Amide **290**, due to its low solubility in dimethylformamide, was reacted in methanol/dimethylformamide using sodium ethoxide as a base.[133] The ansa compounds **295, 296,** and **297** were obtained in 45, 23, and 26% yields, respectively. If, instead of **290**, the more soluble 2,5-dimethoxy-p-phenylenediamine bismethane sulfonamide was cyclized with 1,10-dibromodecane in dimethylformamide at 90° in the presence of potassium carbonate, compound **299** was obtained in 68% yield. Amides **295, 296,** and **299** were hydrolyzed with hydrogen bromide in glacial acetic acid/benzene/phenol, to give compounds **300, 301,** and **302**

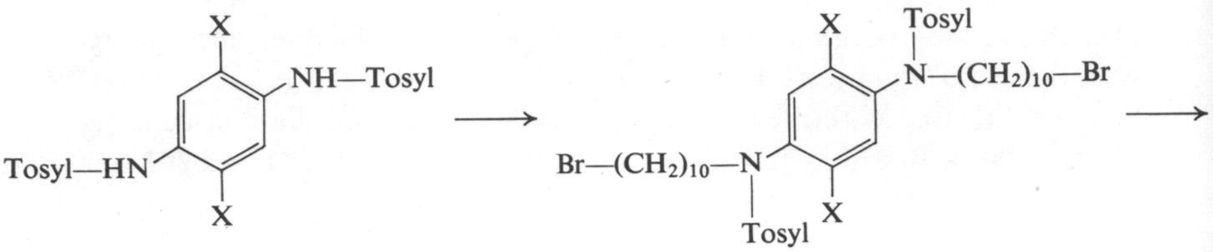

288: X = H
289: X = Br
290: X = OCH$_3$

291: X = H
292: X = Br
293: X = OCH$_3$

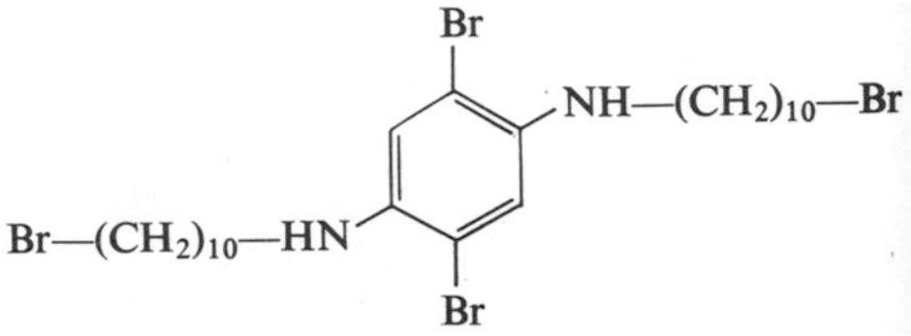

294

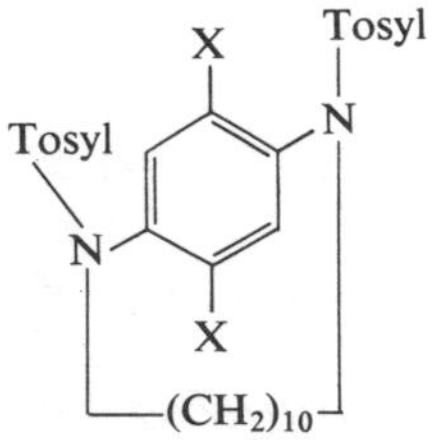

295: X = H
296: X = Br
297: X = OCH$_3$

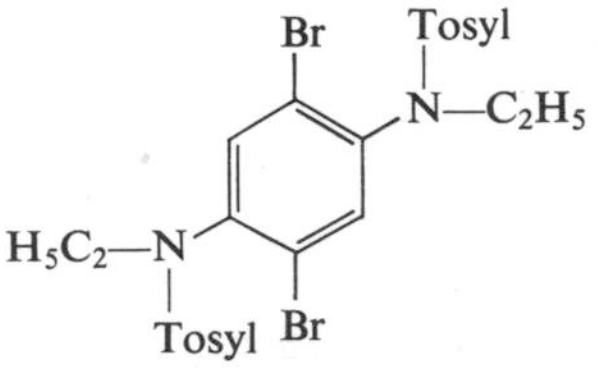

298

in yields of 50, 79, and 80%, respectively. The amines **301** and **302** were also characterized through the N,N-diacetyl derivatives.

Following the method first used by Stetter et al.[90, 139–141] and later by other workers,[142, 143] diamines **301** and **302** were cyclized with sebacic acid dichloride using the high-dilution method in the presence of pyridine.[38] The yields of diansa compounds **303** and **304** were 62 and 78%, respectively. While the dimethoxy compound **304** could be reduced with diborane[128, 144, 145] in tetrahydrofuran or with LiAlH$_4$, to give diansa compound **287a** in 79 and 36–53% yields, no distinct product was obtained with dibromide **303**.

When **302** was cyclized with sebacic acid dichloride and the resulting reaction mixture reduced without carrying out a purification beforehand, small amounts of a dimer could be isolated. The authors tentatively assigned structure **308** to this dimer.

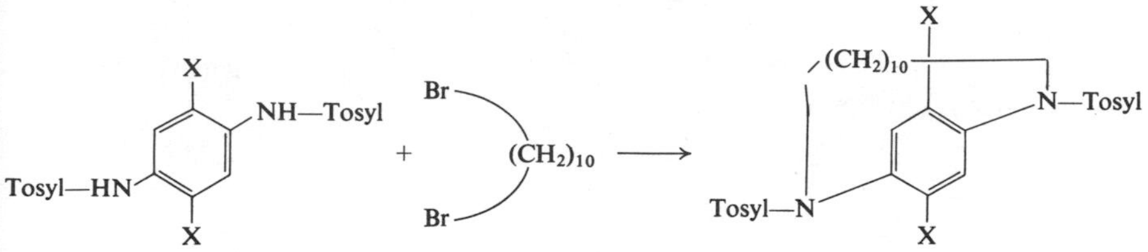

288: X = H
289: X = Br
290: X = OCH$_3$

295: X = H
296: X = Br
297: X = OCH$_3$

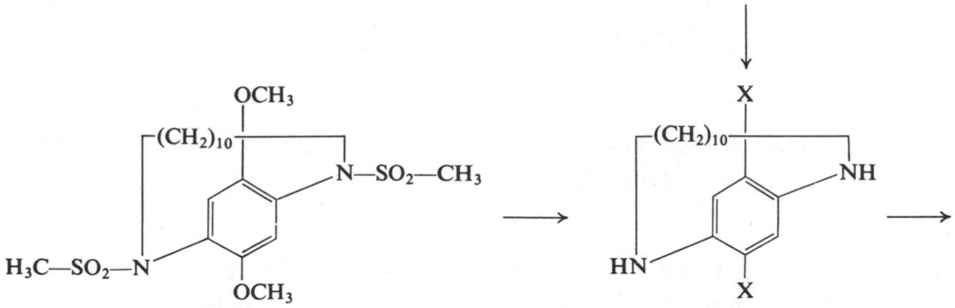

299

300: X = H
301: X = Br
302: X = OCH$_3$

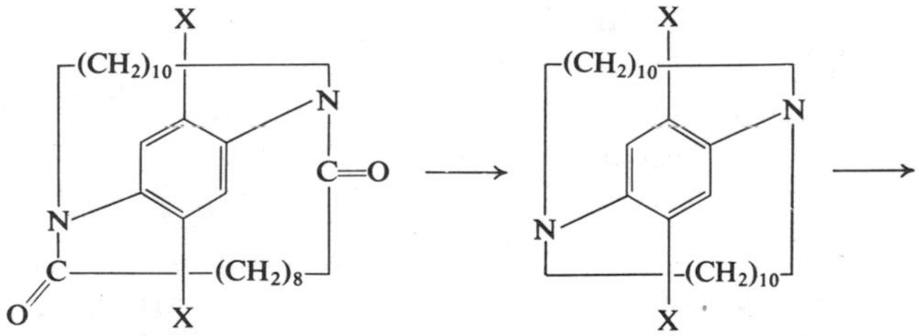

303: X = Br
304: X = OCH$_3$

287a: X = OCH$_3$
305: X = OH
306: X = OAc

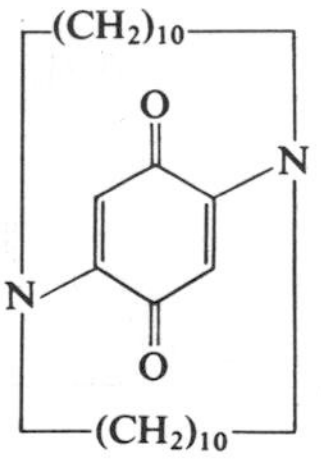

307

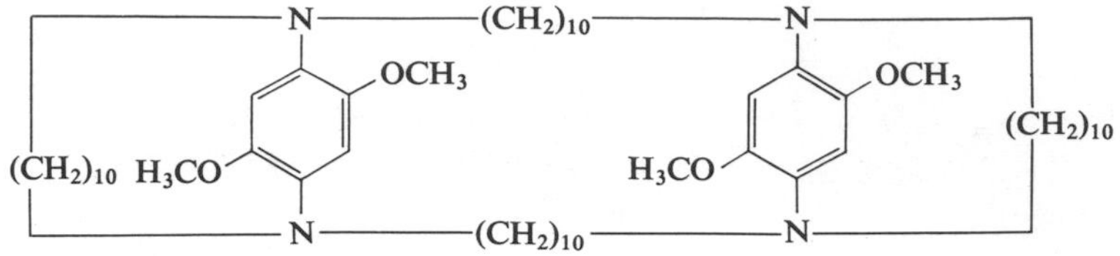

308

Ether cleavage of **287a** and **287b** with AlBr$_3$,[38, 146] or better yet with hydro-iodic acid,[38] resulted in the hydroquinone derivative **305**. This compound in turn was converted by acetylation to **306** and by treatment with ferric(III) salts to benzoquinone **307**.[146]

9.1.2. *Triansa Compounds by the Attachment of Two Additional Bridges to a Hydroquinone Polymethylene Ether*

To obtain a triansa compound, Schill successfully repeated the reaction sequence leading to **287b** using a hydroquinone polymethylene ether of adequate chain length as starting material.[15, 69] In this reaction sequence, however, it was at first uncertain whether the bridge already present would

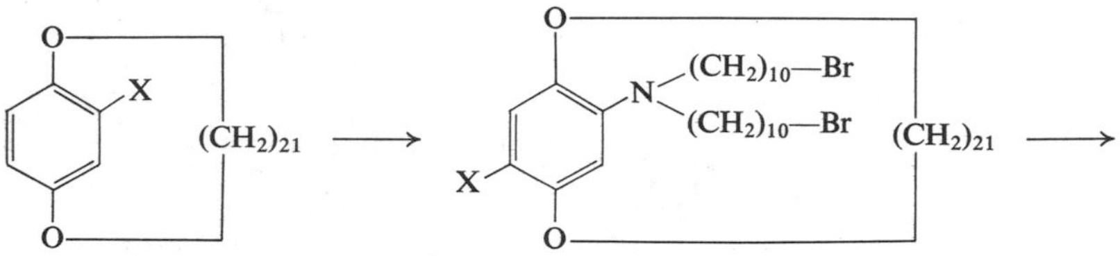

212: X = H
309: X = NO$_2$
310: X = NH$_2$

311: X = H
312: X = NO
313: X = NH$_2$

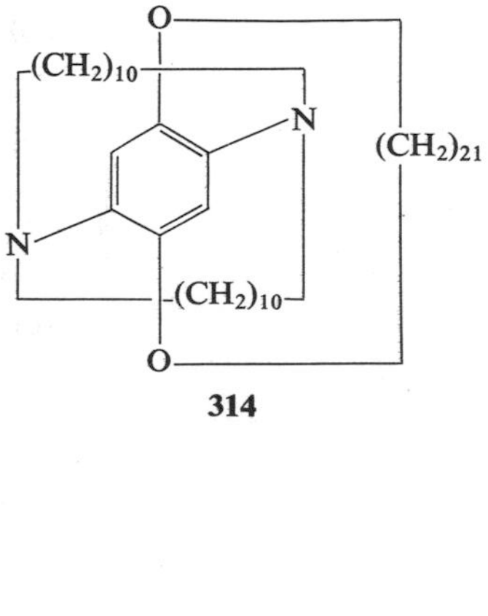

314

shield the aromatic nucleus to such an extent that the attachment of two additional bridges would become impossible.

Using Stuart-Briegleb molecular models, a triansa compound of the type **314** can be built, only if the bridge between the ether oxygen atoms contains at least 18 methylene groups.

From the work on model compounds already discussed, the synthetic path was obvious if a hydroquinone polymethylene ether served as starting material.[15, 69] The hydroquinone polymethylene ether **212** (see Section 8.5) was converted via compounds **309–313**, as described for compound **287b**, to the triansa compound **314** in 2% total yield (based on **310**).

9.1.3. Triansa Compounds from the Cyclization of Bifunctional Diansa Compounds

The synthesis of a diansa compound containing two long chains carrying functional end groups, which would allow the closure of the third bridge,

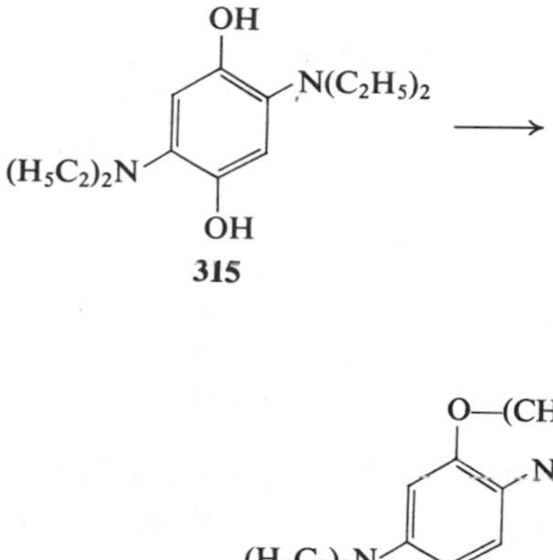

315

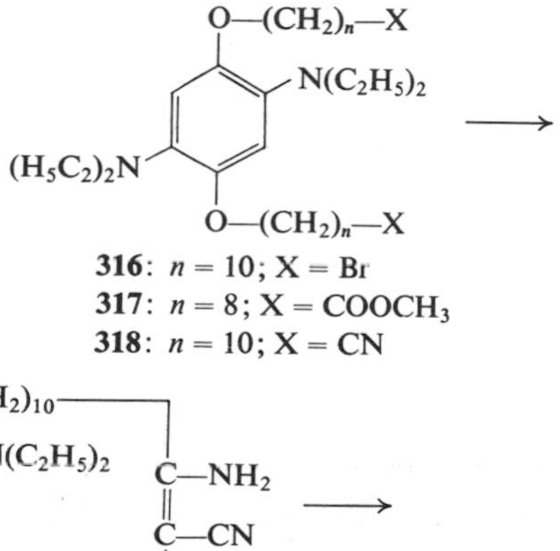

316: $n = 10$; $X = Br$
317: $n = 8$; $X = COOCH_3$
318: $n = 10$; $X = CN$

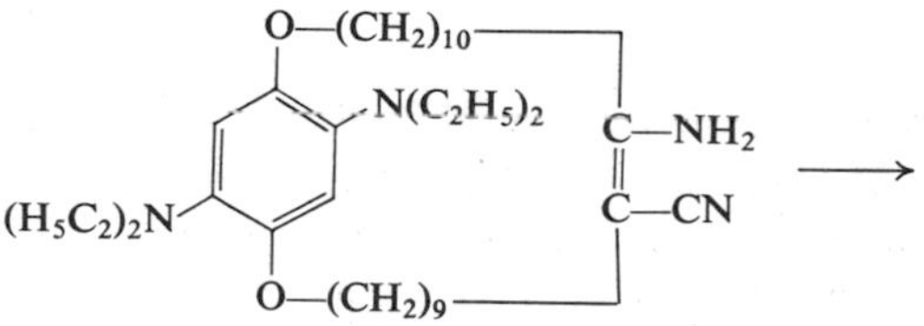

319

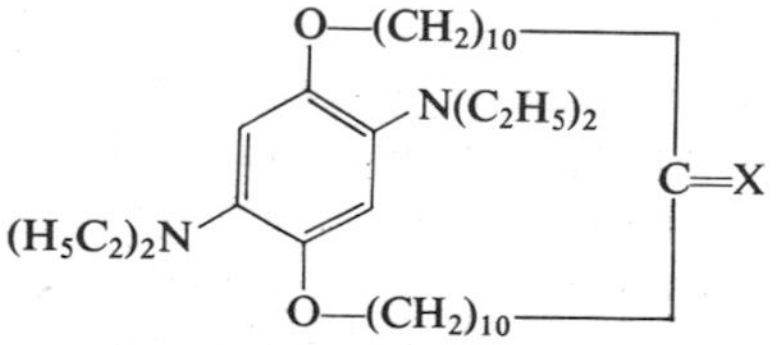

320: $X = O$
321: $X = H_2$

was studied by Doornbos and Strating[38, 131] on the model substance **315**. Its alkylation with various alkyl halides in dimethyl sulfoxide at 20°–25°, using NaH as a base, led to ethers **316, 317,** and **318**. Using other alkylating methods such as methyl iodide in methanolic potassium hydroxide at room temperature, dimethyl sulfate in 4 *N* potassium hydroxide, NaH in benzene at 80°, no reaction could be brought about. Likewise, it was not possible to

322: $n = 8$; $X = COOC_2H_5$
323: $n = 10$; $X = COOC_2H_5$
324: $n = 10$; $X = CN$

325

326: $X = O$
314: $X = H_2$

cyclize the hydroquinone derivative **315** with equimolar amounts of 1,10-dibromodecane. The reaction was carried out in dimethyl sulfoxide under high-dilution conditions with NaH as a base.

Diester **317**, when subjected to an acyloin cyclization gave a compound, which from its IR spectrum should have been the corresponding acyloin. Cyclization of dinitrile **318** according to Ziegler led to the enaminonitrile **319**, which hydrolyzed to **320** and reduced according to Huang-Minlon,[108] gave ansa compound **321** in 33% yield (based on **318**).

In a similar manner as described for **315**, the double-bridge hydroquinone derivative **305** was converted with the corresponding alkyl bromides to the ethers **322, 323,** and **324** in 88, 76, and 96% yields, respectively. Although no reaction occurred when diester **322** was subjected to an acyloin condensation, it was possible to cyclize dinitrile **324**. Ketone **326** was isolated in 70% yield (based on **324**) when the enaminonitrile **325**, obtained from the cyclization, was hydrolyzed in sulfuric acid/glacial acetic acid. Finally, Huang-Minlon reduction led to triansa compound **314**.

The triansa compound **314** synthesized by this path and the compound prepared by Schill by a different path were identical with respect to their IR and UV spectra, melting points, mixed melting points, and thin-layer chromatograms.[38]

Starting from the hydroquinone ether **327** and using a similar method as described for the synthesis of **287b**, Schill[15, 69] was able to prepare, via compounds **328–332**, the diansa compound **333** in 5.5% total yield (based on **327**). With the implementation of the above described synthetic route (see Section 9.1.2), this reaction path was not further pursued.

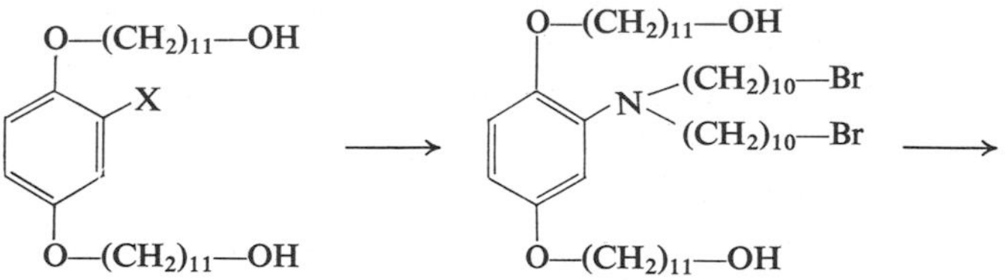

O—(CH₂)₁₁—OH

327: X = H
328: X = NO₂
329: X = NH₂

330: X = H
331: X = NO
332: X = NH₂

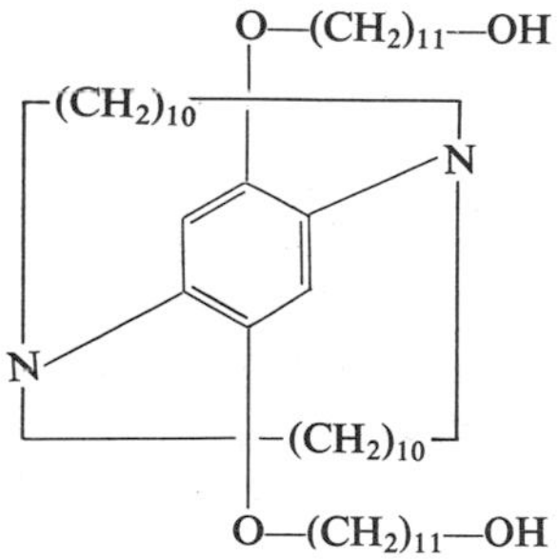

333

9.2. Stereochemistry of Di- and Triansa Compounds

In diansa compounds of *p*-phenylenediamine with a bridge length of 10 methylene groups or more, the unsubstituted benzene ring (see compound

276) is essentially free to rotate inside the double-bridge system. By studying molecular models and from the color of the complex with 1,3,5-trinitrobenzene, Lüttringhaus and Simon concluded that the planar conformation of the ring and the double-bridge system is the preferred geometry of these molecules.[113] For diansa compounds with additional substituents in the 2,5-positions (see compounds 287a, b, and 324) as well as for triansa compound 314, two stereoisomers can be predicted—one in which the nitrogen bonded polymethylene chains are situated on the same side, and one where they are situated on different sides of the benzene ring. Diansa compounds of the type 287a, b can only be separated into their isomers, when the energy barrier between the two forms is increased above a certain minimal value, by ansa bridges which are short enough or by substituents which take up enough space. Since it is likely that the cyclization of the ansa bridges takes place consecutively, it is probable for steric reasons that with a chain length of 10 methylene groups the intraannular stereomer is formed. To gain certainty about the isomeric structure of a diansa compound, its separability into antipodes would have to be tested. If both bridges lie on the same side of the aromatic ring, the molecule has no symmetry elements of the first or second class and should therefore be separable into its antipodes. On the other hand, no separation into antipodes is possible when the ansa bridges are situated on different sides of the benzene ring, since the molecule then possesses a plane of symmetry. In order to be certain about the spatial arrangement, it would be advantageous to first separate the monoansa compound (see compound 302) into its antipodes and thereafter attach the second bridge. Only if the compound hereby obtained is optically inactive can it be concluded that the ansa bridges are on opposite sides of the aromatic nucleus.

Evidence for the presence of the intraannular stereomers of diansa compound 287a and triansa compound 314 was given by the behavior with charge transfer acceptors. Doornbos and Strating[38] showed that, while N,N,N',N'-tetraethyl-1,4-diamino-2,5-dimethoxybenzene gave an intensive colored solution with tetracyanoethylene in chloroform, dichloromethane, ether, and cyclohexane, diansa compound 287a and triansa compound 314 only gave a pale yellow coloration. From this observation it can be concluded that the latter two compounds do not form charge transfer complexes because the aromatic ring is obstructed on both sides by the polymethylene chains.

The dimer of compound 287a, to which structure 308 was assigned, gave a light green solution with tetracyanoethylene in chloroform, dichloromethane, ether, and cyclohexane. It was also observed that while N,N,N',N'-tetraethyl-1,4-diaminobenzene and diansa compound 276 give a colored charge transfer complex with 1,3,5-trinitrobenzene, the substituted diansa compounds **287a**, **308**, the triansa compound **314**, and N,N,N',N'-tetra-ethyl-2,5-dimethoxy-1,4-diaminobenzene do not give a coloration with this substance.

9.3. Investigations of the Cleavage of Triansa Compounds from 2,5-Di-aminohydroquinone Polymethylene Ethers

Under the plausible assumption, that the two macrocyclic systems are linked intraannularly in triansa compound **314**, it would be expected that a rupture of the aryl–nitrogen bonds should result in a catenane. The following methods were considered for this purpose[38, 146]: (1) Birch reduction followed by hydrolysis; (2) electrolysis of quarternary phenylammonium salts; (3) catalytic hydrogenolysis with $Pd/BaSO_4$.

The conversion of the triansa compound to the corresponding benzo-quinone diimmonium salt[147] and especially the subsequent hydrolysis did not seem a promising method, since in water N,N,N',N'-tetramethyl-benzoquinone bisimmonium perchlorate disproportionates to Wurster's blue and an unidentified substance. Benzoquinone is not formed.[148]

9.3.1. *Birch Reduction*

Hydroquinone dimethyl ether as well as N,N-dimethyl aniline can readily be reduced to 2,5-dihydro compounds with sodium in liquid ammonia in the presence of proton donors. The enol ether or enamine thus obtained can be hydrolytically cleaved to cyclohexane-1,4-dione and methanol or cyclohexen-3-one and dimethylamine, respectively.[149]

Diansa compound **287b** was chosen as a model compound. In preliminary experiments the reduction was unsuccessful and the starting material recovered.[146]

9.3.2. *Electrolysis of Quarternary Phenylammonium Salts*

Phenylammonium salts can be cleaved reductively into benzene and tertiary amines with sodium and ethanol in liquid ammonia[150] as well as electrochemically on lead or mercury cathodes. The reaction was discovered by Emmert[151] and extensively studied by Horner and Mentrup.[152]

The quarternation of triansa compound **314**, necessary for the reaction, was again first studied on a model compound.[38, 146] When methyl iodide, dimethyl sulfate, and triethyloxonium tetrafluoroborate were reacted with N,N,N',N'-tetraethyl-2,5-dimethoxy-1,4-diaminobenzene, the corresponding mono- and bismethiodides, the bismethosulfate, and the hexaethyl bis-tetrafluoroborate were obtained. When diansa compound **287a** was reacted with these alkylating reagents, no bisquarternary ammonium salts could be isolated. Only the reaction with methyl iodide at room temperature gave a product which, although it could not be purified was judged to be, from its analysis and NMR spectrum, a monomethiodide.[38] When diansa compound **287b** was warmed for 70 hours at 90° with methyl iodide, a transalkylation resulted in the bisquarternary ammonium salt **334** and 1,10-diiododecane.[146] The failure of the quarternation was attributed to steric hindrances.

Further quarternation reactions of diansa compounds were not attempted. The prospects of a successful cathodic reduction were small anyway, since the ansa bridges are likely to hinder a close approach of the benzene ring to the reaction surface.

334

9.3.3. *Catalytic Hydrogenolysis with Pd/BaSO₄*

According to Kuhn and Haas, some aniline derivatives can be subjected to hydrogenolysis in modest yield with palladium on barium sulfate.[153] As reported by Swaters,[154] cyclohexanone is obtained in 9% yield and diethylamine in 4% yield when N,N-diethylaniline is reduced under these conditions. With diansa compound **287a**, however, no reaction took place.[38]

9.4. Cleavable Diansa Compounds as Model Compounds and as Starting Materials for Catenanes and Rotaxanes

The work described in the preceding chapters showed that a precatenane, such as triansa compound 314, can be synthesized from a diansa compound as well as from a cyclophane. As was seen, the final conversion to a catenane could not be accomplished, however. Therefore, all the subsequent investigations toward the directed synthesis of a catenane were concentrated on finding a triansa compound, which would allow a detachment of the double-bridge system from the aromatic nucleus by a cleavage reaction successful in open-chain systems.[15] The work carried out so far is based on two reactions: (1) hydrolysis of diamino-p-benzoquinones; (2) hydrolysis of amino-o-benzoquinones.

9.4.1. Diansa Compounds from 1,4-Dialkoxy-2,5-diamino-3,6-dialkylbenzene Derivatives

Amino-p-benzoquinones can be hydrolyzed under acidic or basic conditions to the hydroxy-p-benzoquinones and the corresponding amines.[155, 156] The reaction is especially suitable for dialkyl substituted quinones, and often serves as a simple method for obtaining hydroxy-p-benzoquinones. Hence, this method would be useful in a catenane synthesis, if a triansa compound of type 335 could be obtained. Such a triansa compound would require intraannularly linked bridges, in which the outer arches are in the 3- and 6-positions on the aromatic ring. When this compound is cleaved a hydroquinone derivative should be formed, which on dehydrogenation should lead to a diamino-p-benzoquinone derivative 336. On hydrolysis of this compound, catenane 337 should be formed.

The synthesis of a precatenane of type 335 was experimentally attempted by three different paths.[15, 157]:

1. By introducing substituents in the 3- and 6-positions in a diansa compound of the type 287 and then closing the third bridge.

2. By synthesizing a compound of type 287 having substituents in the 3- and 6-positions, and subsequently closing the third bridge by linking these two substrates.

3. By introducing two amino groups in the 2- and 5-positions of a 3,6-polymethylenehydroquinone dialkyl ether, and then linking the two nitrogen atoms via two methylene bridges.

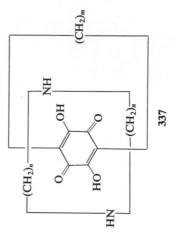

337

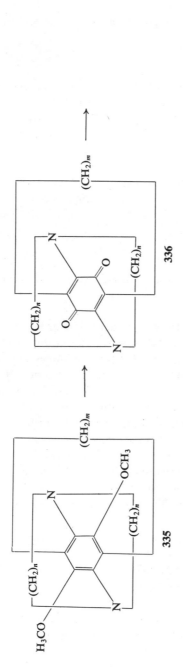

335   ⟶   336   ⟶

9.4.1.1. *Investigations toward the Introduction of Substituents in Diansa Compounds.* The Friedel-Crafts acylation was considered a possible path for the introduction of substituents into diansa compounds of type 287, even though only a few such reactions with *N,N*-disubstituted aromatic amines had been described. In preliminary work *N,N*-diethylaniline was acylated in the 4-position in 3.8 % yield by using acetyl chloride and AlCl₃ in tetra-chloroethane.[38] A similar low yield was obtained in CS₂. On the other hand, when *N,N,N′,N′*-tetraethyl-1,4-diaminobenzene was similarly treated with acetyl chloride or 9-chloroformyl nonanoic acid methyl ester, no reaction took place.

The reaction of halogens with the unsubstituted diansa compound 276 only resulted in a blue Wurster's compound, with no substitution reaction taking place.[113] Since dihalodiaminobenzoquinones lend themselves to various nucleophilic substitution reactions, it was planned to introduce long-chain substrates into the diansa benzoquinone 307, by first substituting the 3- and 6-positions with a halogen.[158] Since no examples for the halogenation of amino benzoquinones could be found in the literature, the reaction was first carried out on the model substances 2,5-bisdiethylamino- and 2,5-bispiperidinobenzoquinones. On chlorination of these compounds in aqueous acetic acid there were obtained, instead of the expected dichloro-diaminobenzoquinones, the 5-membered compounds 338 and 339 which had formed by a ring contraction.[15, 159, 160] Other attempts to halogenate substituted diansa compounds of the type 287 have not been made so far.

$$
\begin{array}{c}
R \\
| \\
HO \quad C{=}O
\end{array}
$$

338: R = N(CH₃)₂

339: R = N⟨hexagon⟩

9.4.1.2. *Synthesis and Cleavage of Diansa Compounds Obtained from 3,6-Disubstituted 1,4-Dialkoxy-2,5-diaminobenzenes.* Since a diansa compound of the type 287 did not permit subsequent introduction of substituents, Schill[146] tried to obtain a diansa compound carrying chlorine in the 3- and 6-positions by a completely different reaction sequence. Chloranil was reacted with 11-bromoundecylamine to quinone **340**, reduced with sodium dithionite to hydroquinone **341** and then converted to the tetraacetate **342**. Hydrolysis of the *O*-acetyl groups led to **343**, which upon methylation with diazomethane gave the *N,N'*-diacetylated diaminohydroquinone dimethyl ether **344**. It now proved impossible to hydrolyze the *N*-acetyl groups. The protection of the amino groups is necessary, however, because **341** cannot be directly methylated to **347**.

340

341: R = R' = H
342: R = R' = COCH$_3$
343: R = H; R' = COCH$_3$
344: R = CH$_3$; R' = COCH$_3$
345: R = H; R' = Z
346: R = CH$_3$; R' = Z
347: R = CH$_3$; R' = H
Z = carbobenzoxy

In a similar reaction path, **341** was converted with carbobenzoxychloride to **345**. Methylation with diazomethane to **346** and cleavage of the protective groups with hydrogen bromide in glacial acetic acid at room temperature, led to the hydrobromide of diamine **347**. Cyclization of this compound in isoamyl alcohol by the high-dilution method in the presence of potassium carbonate did not result in a distinct reaction product. The failure of this reaction was attributed to either steric hindrance or too weak a basicity of the amino groups. The second of these effects, resulting from an inductive effect of the neighboring chlorine atoms, is probably the more important of the two; since a similar compound (see compound **356b**) which had two methyl groups in place of the two chlorine atoms could be cyclized in good yield.

To obtain a diansa compound of the type **287** having alkyl groups in the 3- and 6-positions, Schill and co-workers[15, 157] tried to synthesize a diansa compound of a 2,5-diamino-3,6-dialkylhydroquinone dialkyl ether. The reactions of 2,5-dialkylhydroquinone dialkyl ethers with nitric acid in acetic acid usually gives the corresponding benzoquinones.[161] Exceptions have only been observed with the dimethyl ether of 2,5-bismethoxymethyl-hydroquinone, the dimethyl ether of 2,5-dicyanhydroquinone and with tetrahydrobenzodipyrane. These compounds can be converted to mono- and in the latter case, to dinitro compounds. The different behavior of the hydroquinone ethers is discussed in the original literature.

Dinitrotetrahydrobenzodipyrane **348**, was thought to be a suitable starting material for the synthesis of an alkyl-substituted diansa compound of the type **287**.[15] Compound **348** was reduced to diamine **349**, reacted with 11-chloroundecanoyl chloride to diamide **350** and reduced with diisobutyl aluminum hydride to give diamine **351**. Upon cyclization under the usual conditions, no defined reaction product could be isolated. The reason for the failure of this reaction is uncertain; possibly the reaction ought to be repeated under different conditions.

348

349: R = H
350: R = CO—(CH$_2$)$_{10}$—Cl
351: R = (CH$_2$)$_{11}$—Cl

Since compound **348** could not be used as a starting material for the synthesis of substituted diansa compounds, the dinitrobenzoquinones **352a, c, e** were used. Dinitrobenzoquinones **352a–e** were obtained by treating 2,5-dialkylhydroquinone diesters with nitric acid.[157, 162] In preliminary investigations compounds **352a, c, e** were reduced to the corresponding hydroquinones with sulfurous acid or sodium borohydride and methylated to the ethers **353a, b, d**. Compound **353c** was obtained by reduction of the dimethyl ester **353b** with Ca(BH$_4$)$_2$[163] and subsequent acetylation. Catalytic reduction of **353a, b, c** led to diamines **354a, b, c**. Diamine **354a** was then acylated with 11-chloro- and 11-bromoundecanoyl chloride to **355a** and **355b**, respectively. By reducing these compounds with diborane in tetrahydrofuran,[128, 145] diamines **356a, b** were obtained. Cyclization of **356a** in isoamyl alcohol using the high-dilution method in the presence of potassium carbonate and sodium iodide gave diansa compound **357a** in 58 % yield.

352a: R = CH$_3$
352b: R = n-C$_3$H$_7$
352c: R = (CH$_2$)$_2$—COOH
352d: R = (CH$_2$)$_{11}$—CH$_3$
352e: R = (CH$_2$)$_{12}$—COOCH$_3$

353a: R = CH$_3$
353b: R = (CH$_2$)$_2$—COOCH$_3$
353c: R = (CH$_2$)$_3$—OAc
353d: R = (CH$_2$)$_{12}$—COOCH$_3$

354a: R = CH$_3$
354b: R = (CH$_2$)$_2$—COOCH$_3$
354c: R = (CH$_2$)$_3$—OAc

355a: R = CH$_3$; X = Cl; n = 10
355b: R = CH$_3$; X = Br; n = 10
355c: R = CH$_3$; X = Cl; n = 12
355d: R = (CH$_2$)$_3$—COOCH$_3$; X = Cl; n = 12
355e: R = (CH$_2$)$_3$—OAc; X = Cl; n = 12
355f: R = (CH$_2$)$_3$—OAc; X = Br; n = 12

356a: R = CH$_3$; X = Cl; n = 11
356b: R = CH$_3$; X = Br; n = 11
356c: R = CH$_3$; X = Cl; n = 13
356d: R = (CH$_2$)$_3$—OH; X = Cl; n = 13
356e: R = (CH$_2$)$_3$—OH; X = Br; n = 13

357a: R = CH$_3$; n = 11
357b: R = CH$_3$; n = 13
357c: R = (CH$_2$)$_3$—OH; n = 13

358a: n = 11
358b: n = 13

359

360a: R = H
360b: R = COCH$_3$

With the purpose of later cleaving the aryl–nitrogen bonds, ether **357a** was first hydrolyzed with hydrobromic acid in acetic acid to the hydroquinone derivative and then dehydrogenated with ferric sulfate to diamino-*p*-benzoquinone **358a**. Surprisingly, however, this compound could not be hydrolyzed to the 2,5-dihydroxy-3,6-dimethyl benzoquinone **359** and the corresponding diazamacrocycle. In subsequent model studies it was found that 2,5-dialkyl-3,6-bisdimethylamino-*p*-benzoquinones **361** and **362** were smoothly hydrolyzed in an acid solution. It is remarkable that compound **362**, containing dodecyl substrates, also reacted smoothly.

361: R = CH$_3$
362: R = (CH$_2$)$_{11}$—CH$_3$

These results suggested that the unsuccessful hydrolysis of the diansa compound **358a** was due to a strain in the double-bridge system. It is conceivable that due to the shortness of the methylene bridges, the C-2 carbon atom in the quinoide system can not pass from the trigonal to the intermediate tetrahedral state required for the reaction.

For the reasons just mentioned the homologous diansa benzoquinone **358b**, containing two more methylene groups in each bridge, was synthesized via compounds **354a**, **355c**, **356c**, and **357b**. In this case, a smooth hydrolysis was possible: 2,5-dimethyl-3,6-dihydroxybenzoquinone **359** and macrocycle **360a** were obtained in 90 and 51% yields. The latter was identified by the diacetyl derivative **360b** and, in addition, by a mass spectrum. In an analogous manner starting from diamino compounds **354b** and **354c**, diansa compound **357c** was synthesized via compounds **355d, e, f** and **356d, e** in 8% yield. In the cyclization, both halides **356d, e** gave similar yields. Reduction of diester **355d** could only be accomplished in a stepwise manner. First, Ca(BH$_4$)$_2$ was used to obtain the corresponding alcohol and subsequently diborane in tetrahydrofuran was used to gain **356d**.

It is remarkable that in the cyclization of diamines **356a, c, d, e**, isomeric tricyclic compounds, which would result by the selfalkylation of a chain, are formed in only very small amounts. This is in accordance with the well-known fact that due to transannular effects, intermediate sized rings form only with difficulty; dimers being the preferred products.

Diansa compound **357c** is a suitable starting material for the synthesis of precatenanes and prerotaxanes, provided, that the compound has the intra-annular conformation, i.e., the two bridges lie on opposite sides of the benzene ring. To obtain a precatenane, the two side chains have to be linked by means of a bridge; to obtain a prerotaxane, it is necessary to substitute the two hydroxyl groups on the side chains with bulky substrates. The reaction sequence has not been successful thus far, however, because the corresponding dibromides or ditosylates could not be obtained without an intramolecular reaction taking place at once.

In other cases it was found that the hydroxy groups in the side chains of diansa compounds can be substituted with various functional groups (see Section 9.5). Therefore, the dinitrobenzoquinone **352e**, with two long chains, should be a suitable starting material for the preparation of a precatenane. Another possibility for obtaining a precatenane is the attachment of two additional bridges to a 2,5-polymethylene hydroquinone derivative. The synthesis of the latter compound is reviewed in the following chapter.

9.4.1.3. *Synthesis of a 2,5-Polymethylene Hydroquinone Ether.* To obtain a polymethylene hydroquinone ether with more than 20 methylene groups in the bridge, hydroquinone dimethyl ether was acylated to **363** with 12-(chloroformyl)dodecanoic acid methyl ester via a Friedel-Crafts reaction.[15] The method was based on the procedure of Wasserman and

363: X = O
364: X = H₂

365: X = O
366: X = H₂

367

368

Dawson.[164] Reduction with palladium on charcoal in glacial acetic acid gave **364**. By renewed acylation to **365** and reduction diester **366** was obtained. This compound was cyclized to the acyloin **367** by using the high-dilution method and, without prior purification, reduced via a Clemmensen reduction to 2,5-hexacosamethylene hydroquinone dimethyl ether **368**. After introducing two amino groups, this compound would be a suitable starting material for the synthesis of a precatenane. This reaction sequence, however, has not been carried through so far.

The reactions carried out thus far have demonstrated, that diansa compounds of 1,4-dialkoxy-2,5-diamino-3,6-dialkylbenzene derivatives are suitable systems for the directed synthesis of catenanes and rotaxanes. On one hand, they lend themselves to the necessary preparative procedures required to obtain a triansa compound, and on the other hand, they allow the cleavage of the bonds between the aromatic nucleus and the double-bridge system.

9.4.2. *Diansa Compounds from 5-Aminobenzodioxole and Its Dialkyl Derivatives*

In diansa compounds from 2,5-dimethoxy-3,6-dialkyl-*p*-phenylenedi-amines of the type **357** the two bridges are linked to the aromatic nucleus by two nitrogen atoms. Since an aryl–nitrogen bond can turn about its axis, an extraannular linkage between the aromatic nucleus and the double-bridge system can not be disregarded, even though its occurrence is not very probable.

The intraannularly linked compound certainly becomes the only product, if the double-bridge system is kept as short as possible and if it is fixed, in at least one point, perpendicular to the plain of the aromatic ring. In the search of systems, in which these conditions would be satisfied and which, in a later phase, would allow the conversion into catenanes, 5-amino-benzodioxoles, and 5-amino-6-alkoxybenzodioxoles, as well as their derivatives, were theoretically and experimentally investigated.[15, 165, 166]

For the synthesis of diansa compound **380a**, containing no alkyl substituents on the aromatic ring, it was planned to cyclize the ketal formed between 4-aminocatechol and a ketone consisting of at least 17 carbon atoms. Since in these experiments it was found that long-chain ketones can not be ketalized with catechol derivatives containing electron withdrawing substituents, the ketalization of 1,21-dibromoheneicosan-11-one and catechol to **369a** after the method of Salmi,[93] was carried out before introducing the amino group.

When the ketal **369a** was reacted with nitric acid in glacial acetic acid, a method used by Slooff[167] to nitrate a number of benzodioxoles, a partial hydrolysis of the ketal took place. However, if the reaction was carried out under anhydrous conditions with $Cu(NO_3)_2 \cdot 3\,H_2O$ in acetic anhydride,[109] nitro compound **373a** was obtained in good yield. Reduction with stannous chloride in ether under the addition of hydrogen chloride[167] led to the otherwise not directly obtainable benzodioxole **377a**. Its cyclization in isoamyl alcohol in the presence of potassium carbonate,[113] gave diansa compound **380a** in 24% yield (based on **373a**).

To obtain a triansa compound from **380a** either a polymethylene bridge had to be extended from the 4- to the 6- or 7-position or an analogous reaction sequence with a catechol derivative already containing such a bridge had to be carried out. If the reaction path had been continued with 3,6-dialkyl- or 3,6-polymethylene catechols, the sometimes unstable 2,5-dialkyl 3-hydroxy-*p*-benzoquinones,[168] instead of the stabler 2,6-dialkyl-3-hydroxy-*p*-benzoquinones would be formed as cleavage products (see

369a: R = H; X = Br
370a: R = CH_3; X = Br
371a: R = CH_3; X = Cl
372a, b: R = $(CH_2)_2$—CH_3; X = Cl

373a: R = H; X = Br
374a: R = CH_3; X = Br
375a: R = CH_3; X = Cl
376a, b: R = $(CH_2)_2$—CH_3; X = Cl

377a: R = H; X = Br
378a: R = CH_3; X = Cl
379a, b: R = $(CH_2)_2$—CH_3; X = Cl

380a: R = H
381a: R = CH_3
382a, b: R = $(CH_2)_2$—CH_3

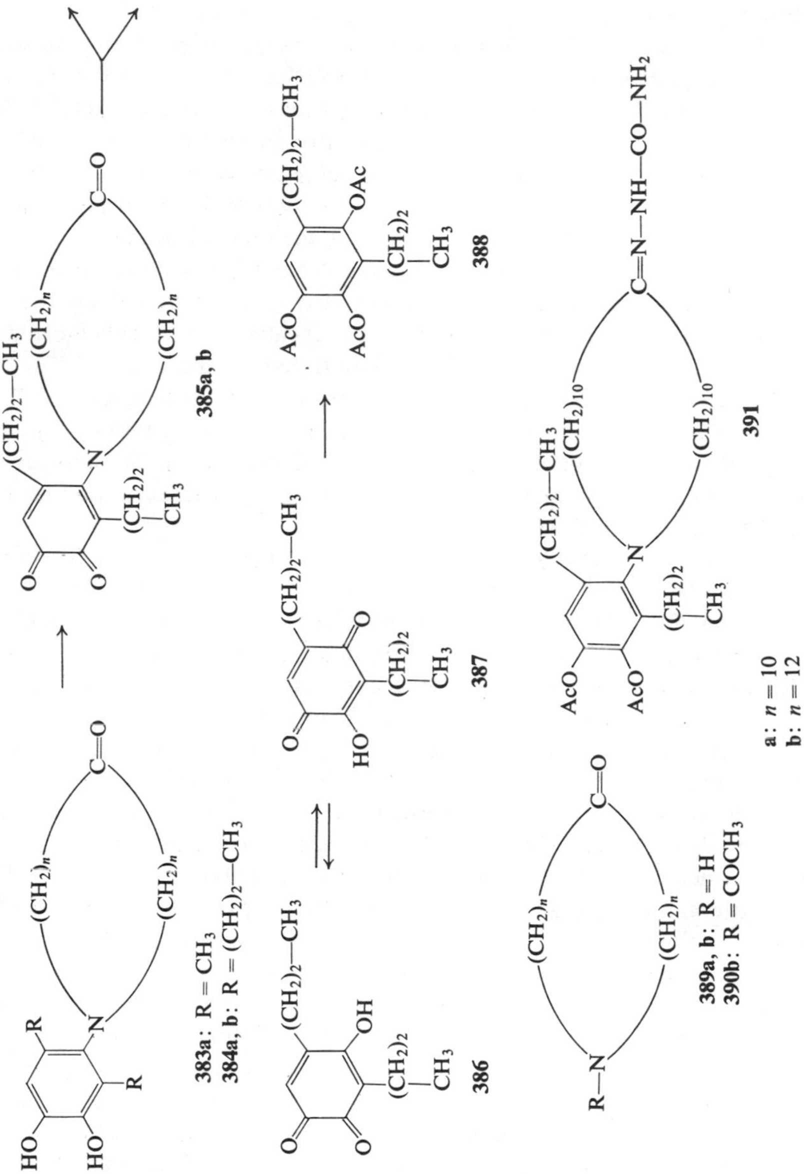

385a, b

383a: R = CH$_3$
384a, b: R = (CH$_2$)$_2$—CH$_3$

386

387

388

389a, b: R = H
390b: R = COCH$_3$

391

a: n = 10
b: n = 12

Section 9.4.2.1). For this reason, the synthesis was continued with 3,5-disubstituted catechols.

The preliminary reactions necessary to elucidate the reaction path were performed on 3,5-dimethyl- and 3,5-dipropylcatechols. The former was reacted, as before, with 1,21-dibromoheneicosan-11-one to ketal **370a**, and nitrated to **374a**. The reduction of the nitro group, the position of which was independently proved, could not be achieved. With stannous chloride solution[167] at room temperature no reaction occurred, and at higher temperature the ketal was cleaved. Catalytic reduction with Raney nickel was not possible, since even under mild conditions reductive dehalogenation takes place.[169] Since aliphatic chlorides are resistant to reduction with Raney nickel, the reaction sequence was repeated with 1,21-dichloroheneicosan-11-one. Benzodioxole **371a** was nitrated to compound **375a** and reduced in isoamyl alcohol with Raney nickel at room temperature. The amine **378a** thus obtained was cyclized in 25% yield to diansa compound **381a** in isoamyl alcohol by the high-dilution method in the presence of potassium carbonate and sodium iodide. Hydrolysis in acid solution led to the aminocatechol **383a**.

A similar reaction sequence with 3,5-di-*n*-propylcatechol gave, via compounds **372a**, **376a**, and **379a**, the diansa compound **382a** in 31% yield (based on **376a**). To cleave this ketal drastic conditions were required. Only after refluxing 36 hours with hydrobromic acid in acetic acid was the cleavage to the aminocatechol **384a** complete. For identification it was acetylated and converted to the semicarbazone **391**.

9.4.2.1. *Detachment of the Macroheterocycle from the Aromatic Nucleus.* The cleavage of the aryl–nitrogen bond in the aminocatechol **384a** could be brought about in two reaction steps: (1) Dehydrogenation of **384a** to the amino-*o*-benzoquinone **385a**. (2) Acid hydrolysis of **385a** to the 22-membered macroheterocycle **389a** and the 3,5-dipropyl-*o*-benzoquinone **386**. The latter tautomerized to the 2-hydroxy-3,5-dipropyl-*p*-benzoquinone **387**. Both reactions could be carried out simultaneously by slightly warming compound **384a** with ferric sulfate in dilute sulfuric acid. The hydroxy-*p*-benzoquinone **387** was reductively acetylated to the triacetoxy benzene derivative **388** in 44% yield (based on **382a**). The macrocyclic azaketone **389a** was also obtained in 44% yield. The use of sulfuric acid in the hydrolysis is to be noted, since with hydrochloric acid side reactions (probably in the form of hydrogen chloride addition to the resulting quinones) occur.

Prior to the above reactions, only few examples of the hydrolysis of amino-*o*-benzoquinones were described[170-173]; in all these cases dihydroxy-*p*-benzoquinones, i.e., very stable quinone derivatives were obtained.

The smooth hydrolysis of **385a** was favored, because it is a *N,N*-dialkyl-4-amino-*o*-benzoquinone in which intramolecular hydrogen bonds between the carbonyl oxygens and the amino group cannot form. This facilitates the hydrolysis of the compound as compared to amino- and *N*-monoalkyl-amino-*o*-benzoquinones.[174]

Due to difficulties (which will be mentioned later) in connection with the detachment of double-bridge systems in a triansa compound, an analogous reaction sequence was carried out with 3,5-dipropylcatechol and 1,25-dichloropentacosan-13-one via compounds **372b**, **376b**, and **379b**. In the cyclization to the diansa compound **382b** a 29% yield (based on **376b**) was obtained. On cleaving the compound under the conditions mentioned earlier, triacetate **388** was obtained in 74% and macroheterocycle **389b** in 61% yield. Acetylation of the latter compound led to the *N*-acetyl derivative **390b**.[16]

9.4.2.2. *Syntheses of Long-Chain 3,5-Dialkyl Catechols and Cyclic 3,5-Polymethylene Catechols.* As shown by the reactions carried out, a triansa compound can be synthesized by (1) cyclizing a diansa compound carrying two long chains having cyclizable end groups, (2) attaching two additional bridges to a simple ansa compound. In both cases it is necessary to synthesize either a polymethylene catechol or a catechol with two alkyl chains of suitable chain length.

Using a Stuart-Briegleb molecular model, a triansa compound can be built from a 3,5-polymethylene catechol, provided that the bridge in the latter consists of at least 14 methylene groups. It seems doubtful, however, that such a bridge could still be obtained by cyclization, and even if this could be accomplished, it is unlikely that the resulting system would still lend itself to an attachment of two additional bridges. For these reasons, syntheses have only been carried out in which the outer bridges consisted of at least 20 methylene groups.

Two methods were considered for the synthesis of long-chain 3,5-dialkyl catechols or 3,5-polymethylene catechols:[15, 107]

1. Two alkyl chains of suitable length having functional end groups are attached to a catechol or catechol derivative and are then cyclized.

    a. Guaiacol and guaiacol derivatives as starting material

    In the presence of BF_3, 6-alkyl guaiacols are acylated in the 4-position with carboxylic acids.[175] The acylation of 4-alkyl guaiacols under the same conditions at low temperatures leads to substitution at the 5-position, at 150°, however, it takes place, under simultaneous ether cleavage, exclusively in the 6-position.[176]

    To synthesize a 6-alkyl guaiacol derivative having a ω-substituted polymethylene chain attached to it, *o*-vanillinbenzyl ether was reacted via a Wittig reaction with excess (10-methoxycarbonyldecylidene)triphenyl-phosporane **393** to ester **394**. The reaction was carried out in dimethyl-formamide according to the method of Bergelson and Shemjakin.[177] The phosphorane **393** was obtained from phosphonium salt **392**[178] and sodium methylate. Catalytic reduction of **394** with Raney nickel and subsequent hydrolysis, led to the guaiacol carboxylic acid **395**, which could be reduced to the alcohol **396** with LiAlH$_4$. Further work on this reaction sequence has not been carried out.

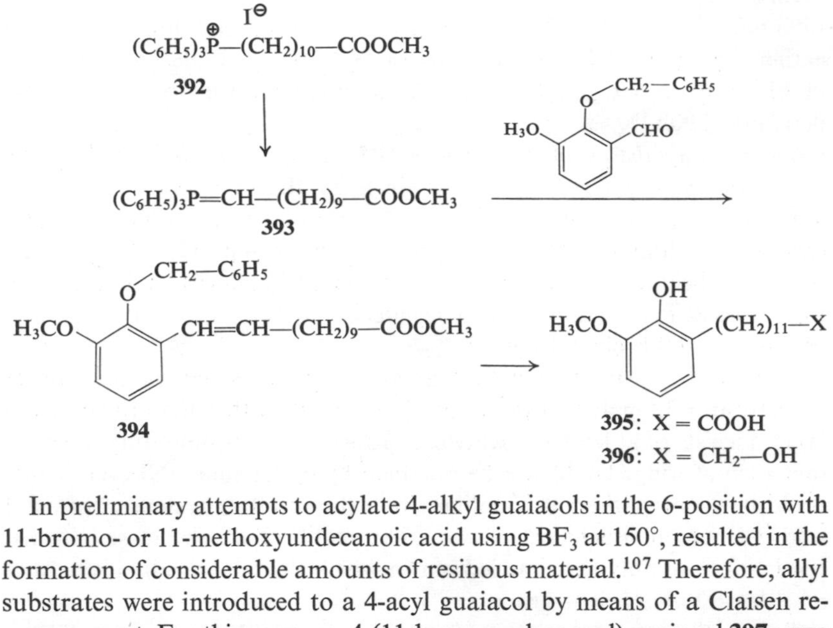

$$\overset{\oplus}{(C_6H_5)_3P}\overset{I^{\ominus}}{}—(CH_2)_{10}—COOCH_3$$

392

$$(C_6H_5)_3P=CH—(CH_2)_9—COOCH_3$$
393

394

395: X = COOH
396: X = CH$_2$—OH

In preliminary attempts to acylate 4-alkyl guaiacols in the 6-position with 11-bromo- or 11-methoxyundecanoic acid using BF$_3$ at 150°, resulted in the formation of considerable amounts of resinous material.[107] Therefore, allyl substrates were introduced to a 4-acyl guaiacol by means of a Claisen re-arrangement. For this purpose, 4-(11-bromoundecanoyl) guaiacol **397a** was prepared by acylation of guaiacol with 11-bromoundecanoic acid in the presence of BF$_3$[175, 179] and then converted to the ether **398a** with allylic bromide. Since this compound decomposed under the conditions of the allylic rearrangement, nitrile **399a** was prepared. Its arrangement led, in 80% yield, to the 4,6-disubstituted guaiacol **400a**, which could be converted to the corresponding veratrole derivative **401a** with dimethyl sulfate and potassium carbonate. A similar reaction with 17-bromoheptanoic acid gave, by the way of **397b–400b**, the veratrole derivative **401b**. As seen from the above reactions it now becomes possible, via a peroxide-catalyzed hydro-gen bromide addition as carried out by Hurt and Hoffman[180] on acyloxyallylbenzenes, to synthesize bifunctional derivatives of 3,5-disubstituted veratroles. More recently Byck and Dawson[181] were able to show that allyl benzenes, when rearranged to the corresponding propenyl derivatives, can be converted in high yield to the aldehydes through ozonolysis and reductive decomposition with dimethyl sulfide.[182]

H_3CO — CO—$(CH_2)_n$—X
RO

397a, b: R = H; X = Br
398a, b: R = CH_2—CH=CH_2; X = Br
399a, b: R = CH_2—CH=CH_2; X = CN

H_3CO — CO—$(CH_2)_n$—CN
RO
CH_2—CH=CH_2

400a, b: R = H
401a, b: R = CH_3

a: n = 10
b: n = 16

b. 4-Hydroxy-5-methoxyisophthalaldehyde as starting material

Starting from the guaiacoldialdehyde **402**, dimethyl ether **403** was obtained by methylation with dimethyl sulfate and potassium carbonate. In a Wittig reaction of this compound with phosphorane **393**, according to the method of Bergelson and Shemjakin,[177] diester **404** was obtained. It was hydrolyzed to the crystalline diacid **405** in 67% yield (based on **403**), catalytically hydrogenated to the diacid **406**, and esterified. Surprisingly, the ester **407** could not be subjected to an acyloin cyclization according to Hansley,[11] Prelog et al.,[12] and Stoll et al.[13]; only starting material was isolated. The failure of the reaction may be attributed to the oxygen atoms in the ether. These may cause such a strong adsorption of the molecule onto the sodium, that the thin layer of the salt of the enediol first formed completely covers the sodium surface and prevents further reaction. A further observation which makes this explanation probable is that the 1,4-dimethoxy-benzenedicarbonic acid dimethyl ester **366** (see Section 9.4.1.3), having a similar structure, can be cyclized by the acyloin method.

An acyloin cyclization was also carried out according to Schräpler and Rühlmann in the presence of trimethyl chlorosilane.[183] This method, which still allows cyclization when the usual procedure fails, gave, after hydrolysis and a Clemmensen reduction, the 3,5-tetracosamethylene veratrole **415** in 10% yield (based on **407**).[85]

Reaction of the acid **406** with thionyl chloride led to the corresponding acid chloride, which after treatment with aqueous ammonia, gave diamide **408**. This compound, which was difficult to purify, was dehydrated with thionyl chloride to the dinitrile **409** and then, without further purification, cyclized by the method of Ziegler et al.[8,9] After hydrolysis and remethyla-

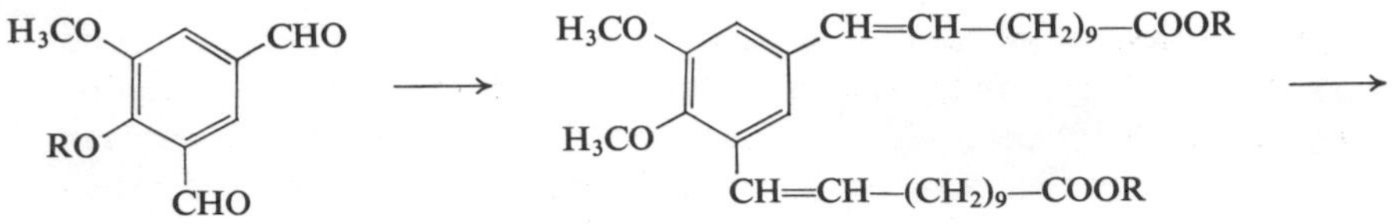

H₃CO⬭CHO → H₃CO⬭CH=CH—(CH₂)₉—COOR →

H_3CO ... CHO

RO ... CHO

H_3CO ... CH=CH—(CH$_2$)$_9$—COOR

402: R = H
403: R = CH₃

404: R = CH₃
405: R = H

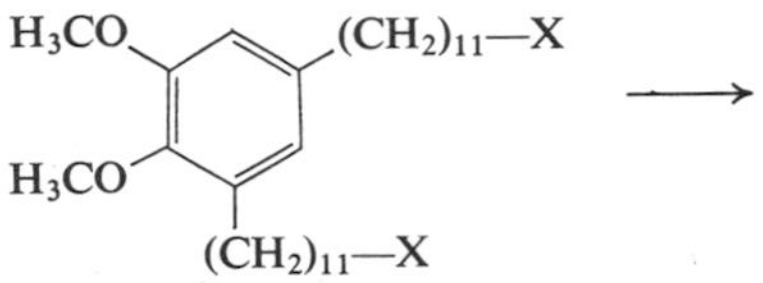

H_3CO ... (CH$_2$)$_{11}$—X

H_3CO ... (CH$_2$)$_{11}$—X

406: X = COOH
407: X = COOCH₃
408: X = CONH₂
409: X = CN
410: X = CH₂—OH
411: X = CH₂—Br
412: X = CH₂—CN

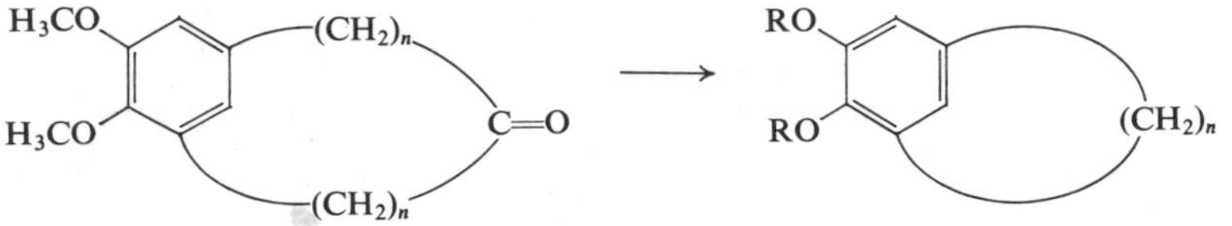

H_3CO ... —(CH$_2$)$_n$ ... C=O ... —(CH$_2$)$_n$

H_3CO

413: n = 11
126: n = 12

RO ... (CH$_2$)$_n$

RO

414: n = 23; R = CH₃
415: n = 24; R = CH₃
127: n = 25; R = CH₃
416: n = 25; R = H

tion of the cyclization product, the cyclic ketone **413** was obtained in 3.8 %
yield (based on **406**).[85]

A higher total yield of a cyclic polymethylene veratrole was obtained[15, 107]
when diester **407** was reduced with LiAlH₄ to diol **410**. The diol was then
reacted with hydrogen bromide, remethylated to dibromide **411**, and converted to the dinitrile **412**. This compound was cyclized according to
Ziegler *et al.* using sodium methylanilide as condensation agent. The
enaminonitrile thus obtained, was hydrolyzed with sulfuric acid in glacial
acetic acid and remethylated to the macrocyclic ketone **126** in 63–76 % yield
(based on **412**). Reduction of this compound by the Huang-Minlon
method[108] gave **127**, as well as partially demethylated products. Complete
cleavage of the ether to the 3,5-pentacosamethylene catechol **416** was

accomplished by using hydrobromic acid in a mixture of acetic acid and phenol.

c. 4-Alkyl veratroles as starting materials

An interesting path toward the synthesis of 3,5-disubstituted veratrole derivatives consists of the metalation of 4-alkyl veratroles with butyllithium, and subsequent alkylation. This method was employed by Byck and
Dawson for the synthesis of 3-alkyl veratroles.[184] The ether cleavage, which
frequently occurs in the synthesis of disubstituted catechol ethers, can be
avoided by use of the corresponding catechol ketals.[185]

2. Syntheses of 2,4-dialkyl phenols or phenol derivatives and subsequent
introduction of a second phenolic hydroxy group.

Another possibility for the synthesis of 3,5-dialkyl catechols or cyclic
3,5-polymethylene catecholes, consists of the introduction of a phenolic
hydroxy group into suitable substituted phenols or phenol derivatives. The
synthesis of the latter compounds is simpler because phenols and phenolic
ethers are easily substituted in the 2- and 4-positions.

In a model reaction it was shown[185] that when 2,4-dimethyl anisole is
reacted with butyllithium, converted to the corresponding Grignard compound with butyl-MgBr, and then treated with air[186] it gives the guaiacol
derivative **417** in 53% yield.

417

To obtain a 2,4-dialkyl anisole containing alkyl substrates which have a
suitable chain length and contain functional end groups, anisole was acylated according to the method of Papa, Schwenk, and Hankin[187] to the keto
ester **418**.[15] The reaction was carried out with 12-(chloroformyl)dodecanoic
methyl ester in the presence of $AlCl_3$ in tetrachloroethane. Hydrolysis of
the ester to acid **419** was followed by the hydrogenation with Pd/C in glacial
acetic acid and resulted in compound **420**. The corresponding methyl ester
421 was again acylated with 12-(chloroformyl)dodecanoic acid methyl ester
to keto diester **422**. Hydrolysis to **423** and catalytic reduction gave the diacid
424, which was converted to dimethyl ester **425**. Further steps in this reaction sequence have not been carried out.

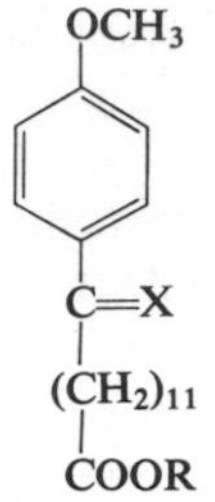

418: R = CH$_3$; X = O
419: R = H; X = O
420: R = H; X = H$_2$
421: R = CH$_3$; X = H$_2$

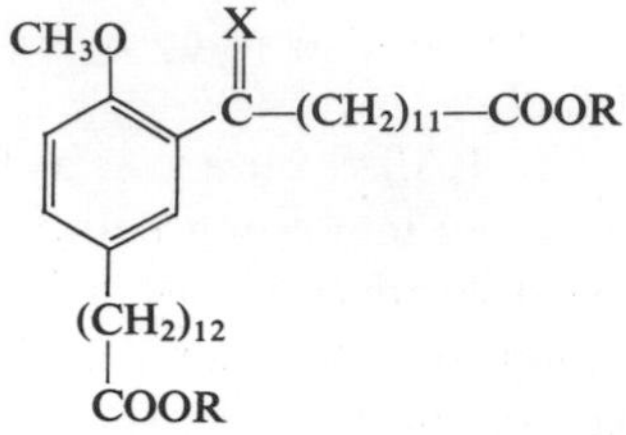

422: R = CH$_3$; X = O
423: R = H; X = O
424: R = H; X = H$_2$
425: R = CH$_3$; X = H$_2$

9.5. Diansa Compounds from 5-Aminobenzodioxoles Having Long-Chain ω-Substituted Alkyl Substrates

The ω-substituted dialkyl catechols, suitable for the synthesis of a diansa compound, should have functional end groups, i.e., they should not interfere in the reactions worked out in the preliminary studies and should be replaceable by cyclizable functions. The hydroxy groups satisfy these requirements.

In model investigations,[25] catechol was acylated with 11-bromoundecanoic acid in the presence of BF$_3$.[175, 179] The bromide 426 thus obtained,

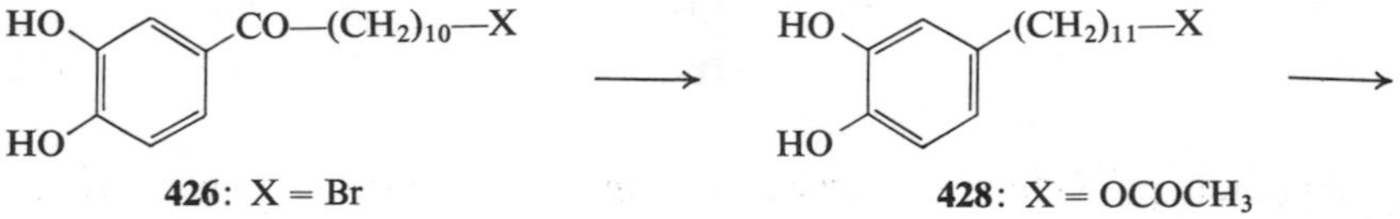

426: X = Br
427: X = OCOCH$_3$

428: X = OCOCH$_3$
429: X = OH

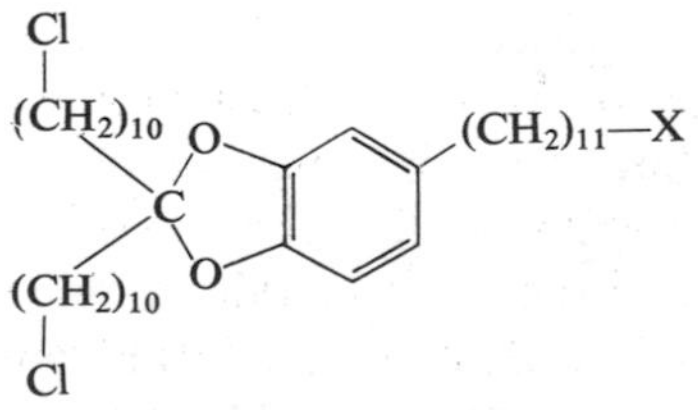

430: X = OCOCH$_3$
431: X = OH

was converted to the acetoxy compound **427**, and then catalytically reduced with Pd/C to the acetoxyalkyl catechol derivative **428**. Hydrolysis gave compound **429**. Ketalization of **428** and **429** with 1,21-dichloroheneicosan-11-one,[91] using *p*-toluenesulfonic acid as catalyst, gave ketals **430** and **431**.

As starting material for the synthesis of diansa compound **439**,[25, 26] the veratrole diacid **406** was used. It was demethylated with hydrobromic acid in acetic acid to the catechol derivative **432**, reduced with LiAlH$_4$ to the catechol diol **433** and ketalized in 87 % yield with 1,21-dichloroheneicosan-11-one to compound **434a**. The diacetate **435a**, obtained from the ketal, was nitrated to **436a** with cupric nitrate in acetic anhydride,[109] saponified to diol **437a** and catalytically reduced to **438a**. Without isolation, this compound was cyclized in isoamyl alcohol using the high-dilution method in the presence of potassium carbonate and sodium iodide. After the acetylation of the reaction product, diansa compound **440a** was chromatographically separated and hydrolyzed to diol **439a**. The yield of the diacetate **440a** was 20 % (based on **437a**).

432: R = COOH
433: R = CH$_2$—OH

434a, b: R = R′ = H
435a, b: R = H; R′ = COCH$_3$
436a, b: R = NO$_2$; R′ = COCH$_3$
437a, b: R = NO$_2$; R′ = H
438a, b: R = NH$_2$; R′ = H

439a, b: R = H   **a**: $n = 10$
440a, b: R = COCH$_3$ **b**: $n = 12$

The cleavage of the ketal linkage in a diansa compound such as **439**, results in a macroheterocycle which is half detached from the aromatic nucleus. Such a cleavage could not be achieved, even under drastic conditions, if the diansa compound contained a 22-membered macrocycle (see Section 9.8).[16] When the number of links in the ring was increased to 26 the hydrolysis proceeded smoothly. Due to these findings, a homologous diansa compound **439b** was synthesized with 1,25-dichloropentacosan-13-one via compounds **434b–438b** and **440b**. At first, a 37% yield of the diansa compound **439b** was obtained (based on **437b**),[25] on refining the experimental procedure, however, the yield could be improved to an average of 65%.[188] This increase was primarily caused by varying the ratio of potassium carbonate to sodium iodide in the cyclization. The structure of some of the side products, obtained in the cyclization of **438a** could also be elucidated; the side products stem primarily from the reaction of the halide with the isoamyl alcohol serving as solvent.[25]

9.6.   Precatenanes from the Cyclization of 5-Aminobenzodioxole Diansa Compounds Having ω-Substituted Long-Chain Alkyl Substrates

As was evident from the work of Doornbos and Strating,[38] triansa compounds can be obtained from diansa compounds having ω-substituted alkyl substrates of appropriate chain length. The reaction steps necessary for the closure of a third bridge on diansa compound **439** were first studied with preliminary reactions on other double-bridge systems.[25, 189]

9.6.1.   *Preliminary Investigations*

Veratrole diol **410** (see Section 9.4.2.2) was chosen as a model substance. It was esterified with methyl sulfonyl chloride to **441** in a 65% yield. The reaction of the diol with PBr₃ or thionyl chloride in benzene in the presence of pyridine did not lead to conclusive results. When reacted with pyrocatechylphosphorus trichloride[190] or tribromide,[191] dihalides **442** and **411** were obtained in 40 and 23% yield, respectively. When diol **410** was reacted with triphenylphosphine and carbon tetrachloride,[192] dichloride **442** was isolated in a 67% yield. The analogous reaction with carbon tetrabromide gave bromide **411** in 58% yield. When the diol was reacted with triphenyl-

phosphine dibromide,[193] the yield was 58 % and when N,N-dimethylaniline
was added, it rose to 79 %.

The cyclization of dibromide 411 with p-toluenesulfonamide in a molar
ratio of 1:1 gave macrocycle 443 in 32.6 % yield. The cyclization was carried
out in dimethylformamide at 120°–130° under high-dilution conditions in
the presence of potassium carbonate. In addition, about 3 % of dimers were
formed which consisted of a mixture of compounds 445 and 446. Detosylation of 443 to compound 444 with sodium in butanol[135] could be accomplished in 72 % yield. It was found that in order to obtain a high yield in this
reaction it is advisable to use an excess of sodium.

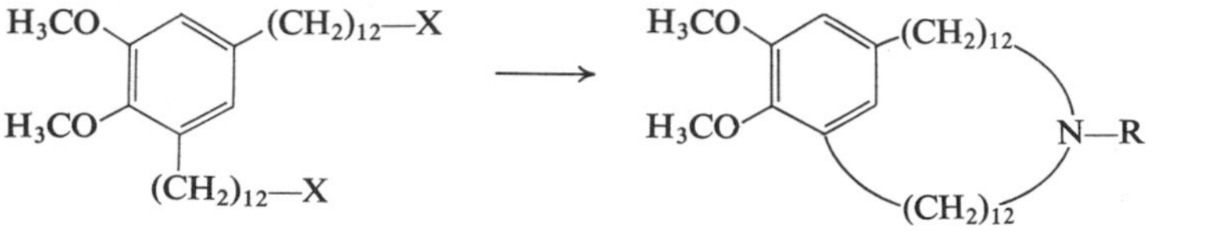

441: X = OSO$_2$CH$_3$
442: X = Cl
411: X = Br

443: R = SO$_2$—C$_6$H$_4$—p—CH$_3$
444: R = H

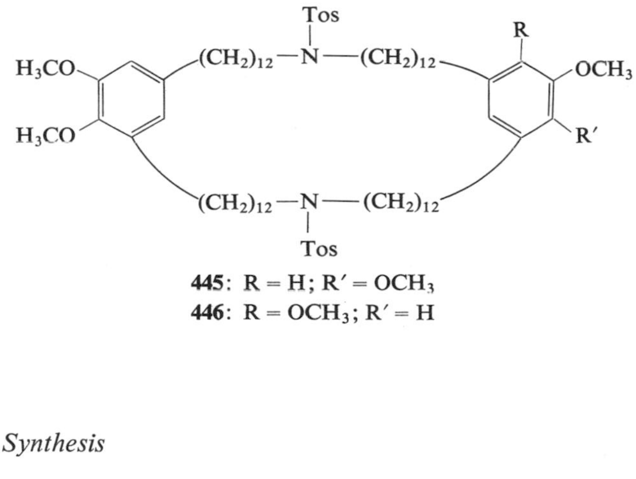

445: R = H; R' = OCH$_3$
446: R = OCH$_3$; R' = H

9.6.2. Synthesis

Dichlorides 447 and 448 were obtained in 37 and 27 % yields, respectively,
by reacting diols 439a, b with pyrocatechylphosphorus trichloride[190] in
chloroform at room temperature.[25] Reaction of 439b with triphenylphosphine and carbon tetrachloride[192] in dimethylformamide or esterification
with methylsulfonyl chloride did not yield a distinct reaction product.

Treatment of diansa compound **439b** with triphenylphosphine dibromide[193] in benzene gave dibromide **449** in 52% yield.[25] This yield could be increased to 91%, however, by improving the reaction conditions.[188] Reaction of the diol **439b** with hydrogen bromide in glacial acetic acid at 90–100° led to the dibromide in 36% yield.

Analogous to the synthesis of **443**, compound **449** was cyclized with *p*-toluenesulfonamide under high-dilution conditions to precatenane **450** in 20% yield. In addition, a mixture of cyclic dimers **519** and **520** (see Section 12.2) was isolated in 2% yield. Detosylation of triansa compound **450** with sodium in butanol[135] gave amine **451** in high yield. Compound **451** was then acetylated to **452**.

447: X = Cl; n = 10
448: X = Cl; n = 12
449: X = Br; n = 12

450: R = SO$_2$—C$_6$H$_4$-*p*-CH$_3$
451: R = H
452: R = COCH$_3$

9.7. Synthesis of Precatenanes by the Attachment of Two Additional Bridges to a 3,5-Polymethylene Catechol

Another possible method of synthesizing a triansa compound consists in the attachment of two additional bridges to a simple ansa compound.[69] In order to carry out such a synthesis,[16] 3,5-pentacosamethylene catechol **416** (see Section 9.4.2.2) was ketalized with 1,21-dichloroheneicosan-11-one to **453a**. Nitration with cupric nitrate in acetic anhydride[109] and subsequent catalytic reduction with Raney nickel gave nitro compound **454a** and amine **455a**, respectively. The amine, which was not isolated, was cyclized in isoamyl alcohol by the Ruggli-Ziegler high-dilution method in the presence of potassium carbonate and sodium iodide. Precatenane **456a** was isolated in 27% yield (based on **454a**). The high yield of the cyclization shows that the attachment of two additional bridges is not hindered to a larger extent by a polymethylene bridge consisting of 25 methylene groups, than by a short alkyl substrate (see Section 9.4.2).

Due to a conformative strain in the double-bridge system, the ketal linkage in triansa compound **456a** could not be hydrolyzed without, at the same time, cleaving a nitrogen to methylene bond. Therefore, the homologous triansa compound **456b** was synthesized in an analogous reaction path, using 1,25-dichloropentacosan-13-one. The reaction sequence via compounds **453b–455b** gave the triansa compound **456b** in an average yield of 29 % (based on **454b**).

In order to avoid the recurrence of the anomalous ketal hydrolysis, as it occurred with triansa compound **456a**, an attempt was first made to synthesize the homolog **456c** of compound **456a** having 16 methylene groups in each of the two ansa bridges. 1,33-Dichlorotritriacosan-17-one was ketalized with the polymethylene catechol **416**, and the resulting compound **453c** nitrated with cupric nitrate in acetic anhydride to **454c**. Cyclization of the corresponding amine **455c**, which had been obtained by catalytic reduction of the nitro compound, by the high-dilution method and with the usual conditions, did not give product **456c** containing doubly alkylated amino groups.

The failure of the reaction was explained by assuming that after the first alkylation, the relatively high mobility of the 16-membered bridge—extending from the ketal linkage to the nitrogen atom—prevents the aromatic nucleus from turning freely through the 25-membered polymethylene bridge. Hence, the monoalkylated amino group is strongly shielded and a second alkylation can not take place. With a shorter bridge length, such a hindrance does not occur because the bridge arches closely around the benzene ring. The given explanation is also in accord with the observation that even with excess butyl or octyl halide, 5-amino-4,6-pentacosamethylene benzodioxole (see Section 9.8.1) can only be mono- but not dialkylated.

453a, b, c: X = H
454a, b, c: X = NO$_2$
455a, b, c: X = NH$_2$

456a, b, c
a: $n = 10$
b: $n = 12$
c: $n = 16$

9.8.   Catenanes from 5-Aminobenzodioxole Triansa Compounds

Triansa compounds **452** and **456a, b** fulfill the requirements for the directed synthesis of a catena compound, that is, they represent intraannularly linked ring systems.[16, 69] Due to the tetrahedral structure of the ketal carbon atom, the two chains linked to it are fixed perpendicularly to the plane of the benzene ring. This geometry, and the relatively short chain length of the ketones used in the ketalizations practically force an intraannular cyclization to the corresponding di- and triansa compounds. The isomeric triansa compounds having an extraannular structure can therefore be neglected in these cases. If on the other hand, 1,33-dichloro-tritriacosan-17-one were used for the ketalization, it could be expected that on cyclizing of **455c** the extraannular as well as the intraannular isomer would be formed. However, since the ring closures to the double-bridge products take place consecutively, it is unlikely for steric reasons, that the second bridge would form on the same side of the aromatic nucleus as the first one. Hence, even in triansa compounds having longer ketal to nitrogen bridges than the compounds synthesized so far, the intraannular species would be favored.

In order to obtain, through ketal hydrolysis, a one-sided detachment of the two macrocyclic systems, triansa compounds **452** and **456a, b** were subjected to the same reactions which had been found to be successful with the model substances.[16, 91] Compounds **452** and **456b** could be hydrolyzed with hydrobromic acid in acetic acid or propionic acid, respectively, to the intraannularly linked catechol derivatives **457** and **458**. Acetylation of **458** led to the crystalline diacetate **459**. With triansa compound **456a**, however, even after refluxing 50 hours, most of the starting material was recovered. Only after very long reaction times or drastic conditions was the ketal linkage broken; it was then, however, accompanied by a cleavage of a nitrogen to methylene bond. This cleavage is unusual, since N,N-diethylaniline is stable under these conditions.

In studies to find new methods for the cleavage of ketals, BBr_3, which can cleave ethers under mild conditions,[194] was reacted with ketal **470**. The reagent gave catechol and the unsaturated compound **471** in high yield.[16] It was also found that BBr_3 does not cleave the alkyl substrate when reacted with N,N-diethylaniline.

The reaction of BBr_3 in benzene with triansa compound **456a** brought about a ketal cleavage, and at the same time, ruptured a carbon–nitrogen bond in the double-bridge system. The unusual course of this ketal hydroly-

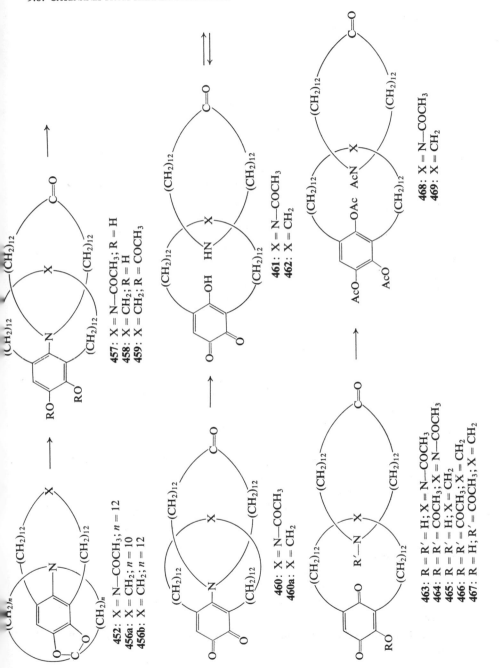

452: $X = N-COCH_3$; $n = 12$
456a: $X = CH_2$; $n = 10$
456b: $X = CH_2$; $n = 12$

457: $X = N-COCH_3$; $R = H$
458: $X = CH_2$; $R = H$
459: $X = CH_2$; $R = COCH_3$

460: $X = N-COCH_3$
460a: $X = CH_2$

461: $X = N-COCH_3$
462: $X = CH_2$

463: $R = R' = H$; $X = N-COCH_3$
464: $R = R' = COCH_3$; $X = N-COCH_3$
465: $R = R' = H$; $X = CH_2$
466: $R = R' = COCH_3$; $X = CH_2$
467: $R = H$; $R' = COCH_3$; $X = CH_2$

468: $X = N-COCH_3$
469: $X = CH_2$

$H_3C\!-\!(CH_2)_9\!-\!C\!-\!(CH_2)_9\!-\!CH_3$

$H_3C\!-\!(CH_2)_9\!-\!C\!\!=\!\!CH\!-\!(CH_2)_8\!-\!CH_3$

Br

470   **471**

sis is attributed to conformative strain in the double-bridge system, even though this is not apparent on a Stuart-Briegleb molecular model. A second, less likely explanation for the results in the hydrolysis of the ketal with hydrobromic acid is that the equilibrium between compound **456a** and the hydrolysis product lies almost completely on the side of the ketal.

Dehydrogenation of amino catechols **457** and **458** to amino-o-benzoquinones **460** and **460a** using ferric sulfate and subsequent acid hydrolysis gave the catena compounds **461** and **462**. The latter spontaneously tautomerize to the p-quinoid structures **463** and **465** and show, due to formation of an inner ammonium salt, the violet color in neutral or basic solution that is typical for 3-hydroxy-2,6-dialkyl-p-benzoquinones.[195]

The postulated structures of catenanes **463** and **465** were verified by physical data as well as chemical reactions. The oily quinones were difficult to purify because they easily decomposed. Their acetylation with acetic anhydride and sodium acetate led to compounds **464** and **466**. Reductive acetylation with zinc powder in acetic anhydride in the presence of triethylamine resulted in the tetra- and pentaacetates **469** and **468** which could be freed from by-products by chromatography. Compound **469**, which was obtained in 76% yield (based on **456b**), was at first oily, but solidified later to crystals with a melting point of 66–72°. It has not yet been possible to recrystallize the compound from solvents, since an oil is always recovered. Basic hydrolysis of **469** and subsequent dehydrogenation with ferric sulfate led to catena compound **467**.

In addition of the structural proof which is contained in the synthesis as such, the following analytical data confirm the structure of catenane **469** as well:

1. A satisfactory elemental analysis was given.

2. The compound gave the expected molecular weight as determined by the isothermal distillation method.[196]

3. The IR and NMR[188] spectra were in accord with the postulated structure.

4. The chromatographic behavior differed clearly from that of its components.

5. The fragmentation pattern of the compound shown by a mass spectrum can not leave any doubt about the formulated structure (see Section 9.10).
With the use of similar arguments the structure of catenane **468** was confirmed.[188]

9.8.1. Synthesis of Molecular Subunits of Catenanes

A synthesis of one of the molecular subunits of catenanes **467**, **468**, and **469**, that is N-acetyl-14-azacyclohexacosan-1-one **390b** was already discussed in Section 9.4.2. Another synthesis of heterocyclic ketones of the type **390b** was carried out by Schill and Zürcher.[188] Cyclization of ketal **369a** with p-toluenesulfonamide in a 1:1 ratio in the presence of potassium carbonate and using dimethylformamide as solvent, gave the cyclic tosylamide **472** in 20% yield. Detosylation and acetylation to **473**, followed by ketal cleavage, resulted in the macrocyclic azaketone **474**. The synthesis

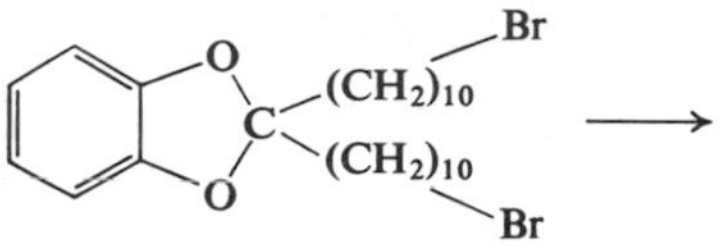

369a

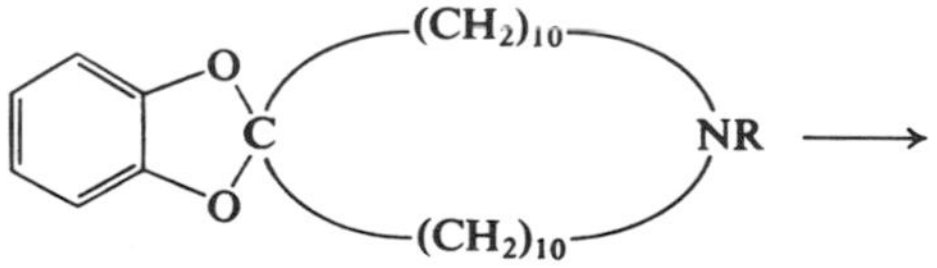

472: R = O_2S—C_6H_4-p-CH_3
473: R = $COCH_3$

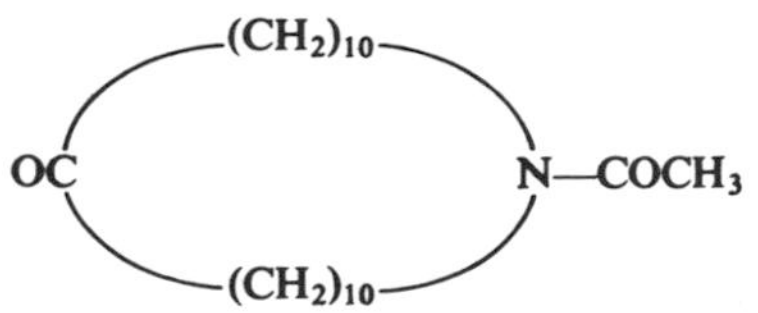

474

of the other two components of catenanes **467** and **469**, that is, 2-hydroxy-3,5-pentacosamethylene-p-benzoquinone **142** and 1,2,4-triacetoxy-3,5-pentacosamethylene benzene **145** was achieved as 3,5-pentacosamethylene catechol **416** was ketalized with diethyl ketone to **475**. Nitration follows:[16]

to **132** and subsequent catalytic reduction led to **476**. This amine was alkylated with *n*-butyl bromide and *n*-octyl chloride to the monoalkyl derivatives **477** and **480**. The reactions were carried out in isoamyl alcohol and in the presence of potassium carbonate and sodium iodide. Even if an excess of halides was used, mainly the monoalkylated products were obtained. Without isolation, the two amines were then acetylated to compounds **478** and **481**, reduced with LiAlH$_4$ to amines **479** and **482**, and hydrolyzed to the amino catechols **483** and **484**. Dehydrogenation of these compounds to the corresponding *N,N*-dialkylamino-*o*-benzoquinones and subsequent acid hydrolysis gave the macrocyclic hydroxybenzoquinone **142**. Compound **145** was obtained by reductive acetylation with zinc powder and acetic anhydride in the presence of triethylamine.

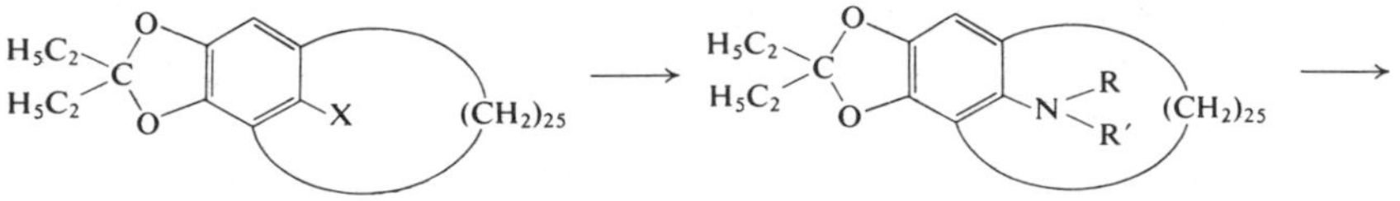

475: X = H
132: X = NO$_2$
476: X = NH$_2$

477: R = H; R' = *n*-C$_4$H$_9$
478: R = COCH$_3$; R' = *n*-C$_4$H$_9$
479: R = C$_2$H$_5$; R' = *n*-C$_4$H$_9$
480: R = H; R' = *n*-C$_8$H$_{17}$
481: R = COCH$_3$; R' = *n*-C$_8$H$_{17}$
482: R = C$_2$H$_5$; R' = *n*-C$_8$H$_{17}$

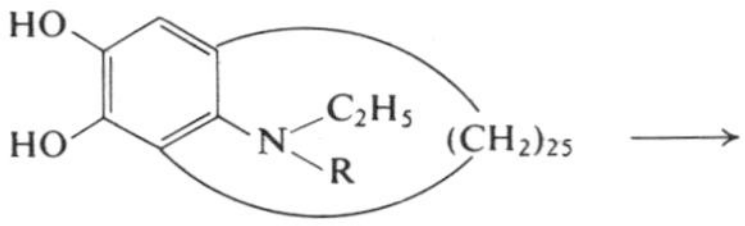

483: R = *n*-C$_4$H$_9$
484: R = *n*-C$_8$H$_{17}$

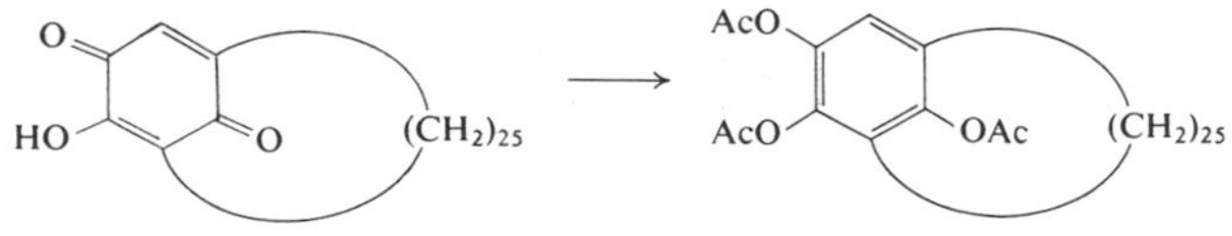

142

145

9.9. Diansa Compounds from 4,7-Dialkyl-5-amino-6-methoxybenzo-dioxoles

The diansa compounds of 5-aminobenzodioxoles represent important intermediates in the directed synthesis of catenanes and rotaxanes. Having in mind the synthesis of oligomeric compounds of this class, Schill and Henschel[166] investigated the possibility of using 2,5-dialkyl hydroquinones as starting materials instead of the 3,5-dialkyl catechols which are often difficult to obtain. At the same time, the double-bridge system was to be, contrary to diansa compounds of type **357** (see Section 9.4.1) perpendicularly fixed at the junction to the aromatic ring by means of a ketal linkage.

To work out the individual reaction steps, 1,2,4-trihydroxy-3,6-dimethylbenzene, easily obtained by the Thiele-Winter reaction and subsequent hydrolysis of 2,5-dimethylbenzoquinone, was used as the starting material. It was reacted with cyclohexanone to ketal **485a**, and methylated to **486a**. The nitration with cupric nitrate in acetic anhydride[109] caused extensive ketal cleavage, so that the nitroketal **489a** was only obtained in 17% yield. This result agrees with the observation that with nitric acid 2,5-dialkyl-hydroquinone ethers are converted to the corresponding benzoquinones.[161] After acetylation of **485a** to **487a**, the nitration to **490a** took place largely without ketal cleavage. Hydrolysis to **488a** and subsequent methylation resulted in the nitro derivative **489a**, which could be obtained by the direct method only in poor yield. This preliminary investigation demonstrated that 1,2,4-trihydroxy-3,6-dialkylbenzenes can be ketalized and subsequently nitrated in the aromatic ring, without ketal cleavage taking place.

In an analogous procedure, 1,2,4-trihydroxy-3,6-dimethylbenzene was reacted with 1,21-dichloroheneicosan-11-one to ketal **485b**. As before, the protection of the hydroxy group by an acetyl group rather than a methyl ether group proved to be advantageous. While, via the compounds **487b**, **490b**, and **488b** the nitro compound **489b** was obtained in 54% yield, the methyl ether **486b** only gave 34% yield in both steps. Amine **491**, which was obtained by a catalytic hydrogenation of **489b** in isoamyl alcohol, was cyclized, without previous isolation and in the same solvent, under high dilution conditions in the presence of potassium carbonate and sodium iodide. From this cyclization, the diansa compound **492** was isolated in 30% yield (based on **489b**).

The double-bridge system in compound **492** was partially detached from the aromatic nucleus with hydrobromic acid in acetic acid. Compound **494** thereby obtained, was treated in acid solution with ferric sulfate and led, via the quinoid structure which was not isolated, to dihydroxybenzoquinone **496** and macroheterocycle **389a** in 44 and 54% yields, respectively.

If the ketal hydrolysis was carried out with hydrochloric acid in acetic acid instead of hydrobromic acid, the methoxy group remained largely intact and compound **493** was obtained. The hydrolysis of the N-aryl bond,

performed as mentioned above, gave the 2-methoxy-5-hydroxybenzo-
quinone **495** in 54% yield and the heterocyclic ketone **389a** in 51% yield.

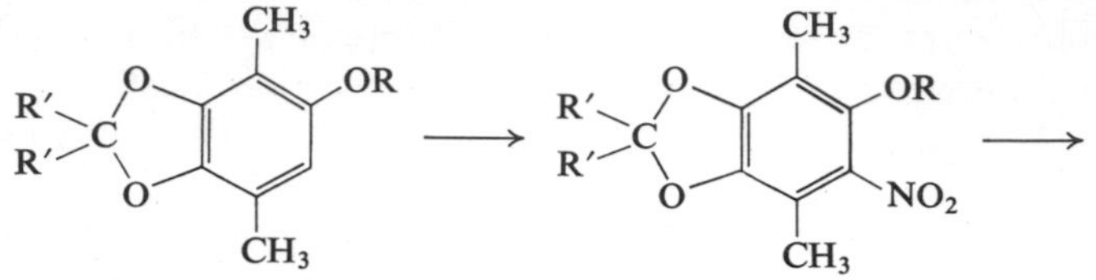

485a, b: R = H
486a, b: R = CH$_3$
487a, b: R = COCH$_3$

488a, b: R = H
489a, b: R = CH$_3$
490a, b: R = COCH$_3$

a: R'—R' = —(CH$_2$)$_5$—
b: R' = (CH$_2$)$_{10}$—Cl

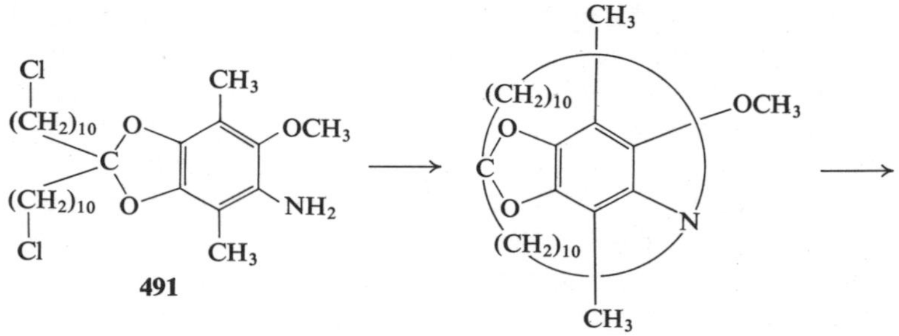

491

492

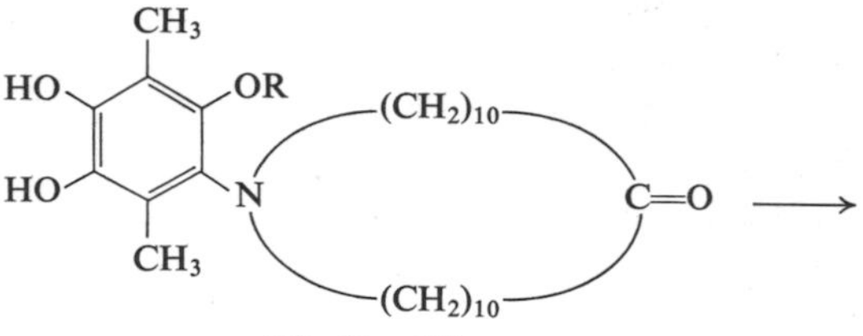

493: R = CH$_3$
494: R = H

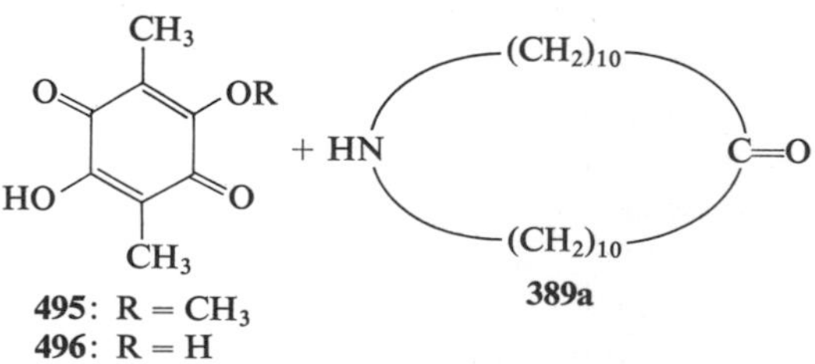

495: R = CH$_3$
496: R = H

389a

10

The Mass
Spectrum of
Catenanes

According to Vetter and Schill,[19] the mass spectrum of a catenane has to fulfill the following requirement: The mass number of the molecular ion peak has to be equal to the sum of the molecular weights of the catenane components. The number of hydrogen atoms in a catenane has to be larger by two than the number which can be calculated from the atoms as well as double bonds and rings present. This requirement arises from the fact that the two macrocycles are joined mechanically and not chemically, and that therefore, no valences are used for the topological linkage.

The fact that the rupture of one of the rings releases the other ring, and therefore destroys the entire catenane structure as such, is expected to influence the fragmentation pattern as follows: All peaks from the components of a catenane which arise while both rings are still closed, must be shifted upwards in the spectrum by the molecular weight of the other ring, while peaks which arise after one of the rings has been opened will remain at their normal positions. This only holds true, however, if one assumes that the ionized subunits do not react with each other.

The intensity of the molecular ion peak of a catenane may be considerably decreased due to its particular structure. In normal cyclic molecules, the molecular ion peak arises not only from those ions which still have the same structure as the un-ionized molecule, but also from those ions in which the ring has already been opened. In the case of a catenated compound, however, only the original cyclic molecule where none of the rings has been ruptured will contribute to the molecular ion peak; while the opened species will appear at the molecular ion peaks of the individual subunits.

In a detailed analysis of the mass spectrum of catenane **469** (see Section 9.8) Vetter and Schill[19] demonstrated that the spectrum was in accord with the predicted behavior of a catenane and thus represented a proof of the catenane structure. An additional peak, which appears in the region of the independent ring masses, is attributed to an ion which arises by a hydrogen transfer from one ring to the other. On the basis of the present evidence it is impossible to assign a definite structure to this ion.

The mass spectrum of catenane **156** (see Section 8.2), a homolog with twice five methylene groups more in the heterocycle, has a mass spectrum which is completely analogous to that of compound **469**.[20] Again, a M + 1 peak of the heterocycle appears in the spectrum, although its intensity as compared to the corresponding peak in compound **469** is substantially decreased.[21] This may be due to the enlargement of the heterocycle by 10 methylene groups which causes the interaction between the two components due to ring size to be decreased.

[3]-Catenanes

11

Methods for Synthesis of [3]-Catenanes

For the synthesis of [3]-catenanes, the statistical method does not need to be considered; the probability that a system consisting of three interlocked rings be formed by purely statistical interactions is extremely small.

A [3]-catenane can be built with Fisher-Hirschfelder molecular models provided that both outer rings contain at least 20 and the middle ring at least 26 carbon atoms. A [2]-catenane, with a winding number $\alpha = 2$, requires either two rings of 33 carbon atoms each or one smaller ring with 30 and one larger with 37 carbon atoms.[3]

On the basis of the reactions worked out for the synthesis of [2]-catenanes, two basic procedures can be considered for a directed synthesis of a [3]-catenane[185, 188, 189]:

11.1. Procedure 1

A diansa compound **497** is synthesized, which carries in the 1,3-positions chains of suitable length with functional end groups X. Dimerization of this compound gives **498**, which upon cleavage of the bonds between the aromatic nucleus and the double-bridge systems, leads to the [3]-catenane **499**.

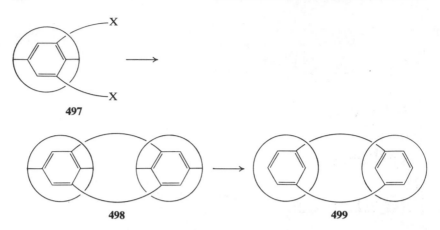

497

498   **499**

A uniform product will only be obtained, however, if compound **497** has
a plane of symmetry perpendicular to the benzene rings. If this is not the
case, as for example in compound **500**, two isomers **501** and **502** may arise.
The formation of such isomers can be avoided, however, if a compound
corresponding to **500** can be synthesized in which the chains of unequal
length carry different functional groups X and Y. If these functional groups
are of such a nature that on dimerization X only reacts with Y, a single
macrocycle **501** is obtained from the unsymmetrical starting material.

In order to suppress the intramolecular cyclization, which competes
with the intermolecular dimerization, the lengths of the chains in com-

501

500

502

pounds of the type **497** are kept as short as possible. On a Stuart-Briegleb molecular model of this compound it can be seen, that a chain length of 5–8 members is most suited.

If by the method discussed above, two structures of the type **503** are dimerized to compound **504**, the synthesis of a [3]-rotaxane **505** becomes possible (R = bulky end groups).

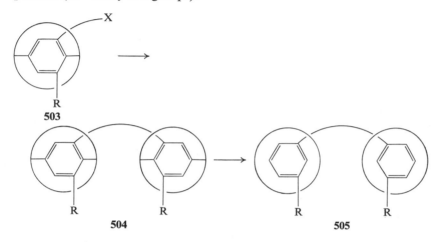

11.2.   Procedure 2

Another possibility to obtain a [3]-catenane **499** consists in attaching two double-bridge systems to a macrocycle **506**, and subsequently cleaving the linkages between the three-ring systems in compound **498** to give **499**.

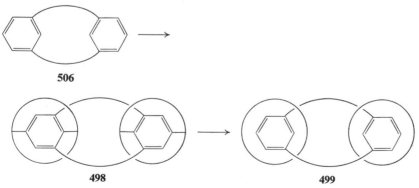

12

Investigations of the Synthesis of [3]-Catenanes according to Procedure 1

12.1. Preliminary Investigations

The synthesis of a [3]-catenane according to Procedure 1 requires the dimerization of a diansa compound having two chains with functional end groups. In preliminary investigations,[188] the veratrole derivative **411** was converted to compound **507** by treating it in dimethylformamide with *p*-toluenesulfonamide and using sodium hydride as a base. In a side reaction small amounts of the macrocyclic compound **443** (see Section 9.6), already mentioned with regard to the cyclization of veratrole dibromide **411** with *p*-toluenesulfonamide, were obtained. The reaction of the veratrole dibromide **411** with **507** in a molar ratio of 1:1, using the high-dilution method in dimethylformamide at 120–130°, resulted in 20–30% yield of a mixture of compounds **445** and **446**. For statistical reasons, one can expect that the compounds are formed in approximately equal amounts. The different positions of the methoxyl groups in the two compounds have little influence on their chemical properties. Chromatographic separation of the two compounds was not possible. Detosylation with sodium in butanol[135] led to a mixture of the two macroheterocycles **508** and **509**.

411   +   507

445: R′ = OCH₃; R″ = H; R‴ = Tosyl
446: R′ = H; R″ = OCH₃; R‴ = Tosyl
508: R′ = OCH₃; R″ = R‴ = H
509: R′ = H; R″ = OCH₃; R‴ = H

12.2. Synthesis

Diansa compound **449** (see Section 9.6.2) represented a suitable starting material for the synthesis of a [3]-precatenane.[188, 197] Reaction of this compound with *p*-toluenesulfonamide in dimethylformamide using NaH as a base gave diamide **512**. Even though excess *p*-toluenesulfonamide was used in all reactions, a considerable amount of partially reacted product was obtained. It probably consisted of a mixture of isomers **510** and **511**. In

510: R = Br; R′ = NH—Tosyl
511: R = NH—Tosyl; R′ = Br

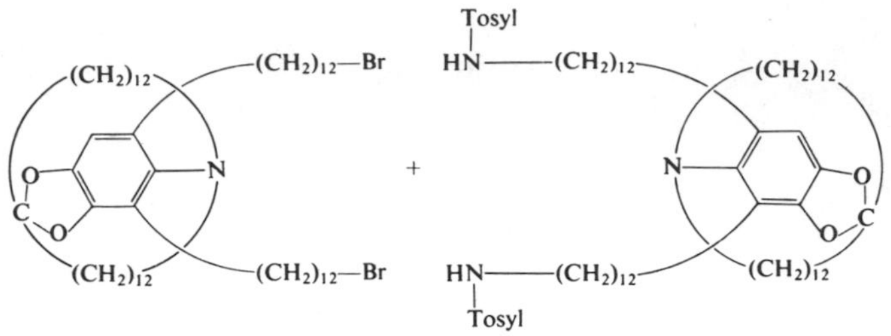

449   **512**

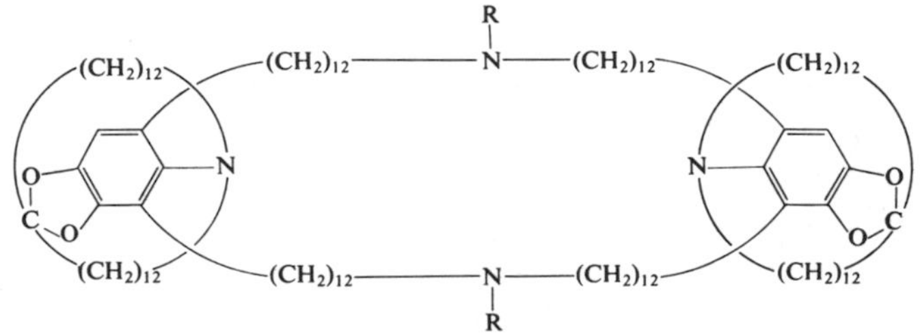

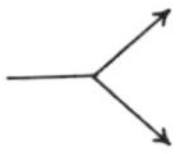

513: R = Tosyl
514: R = H
515: R = COCH$_3$

+

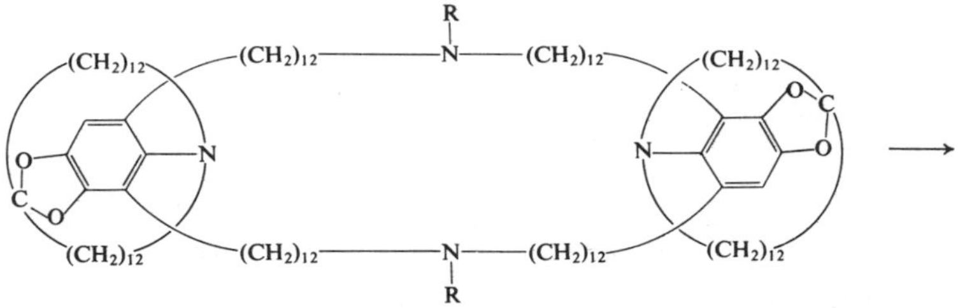

516: R = Tosyl
517: R = H
518: R = COCH$_3$

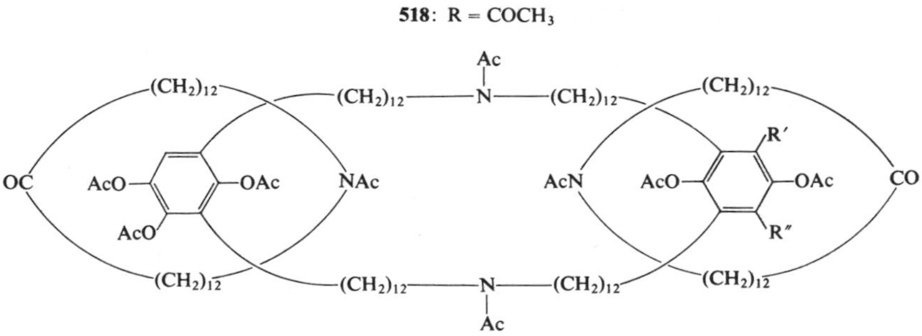

519: R' = OCOCH$_3$; R'' = H
520: R' = H; R'' = OCOCH$_3$

addition, once again the triansa compound **450** (see Section 9.6.2) was formed.

Cyclization of the bistoluenesulfonamide **512** with veratrole dibromide **449** in a 1:1 molar ratio gave a mixture of isomers **513** and **516** in 38 % yield. The cyclization was carried out in dimethylformamide at 120–130° in the presence of potassium carbonate under high-dilution conditions. The mixture of isomers thus obtained, was isolated by chromatography from the other products formed, but could not be separated into its components. Detosylation with sodium in butanol[135] resulted in amines **514** and **517**, which were then acetylated to amides **515** and **518**. The structure of these compounds was confirmed by a mass spectrum; it showed the peak of the molecular ion as base peak, and a fragmentation pattern similar to that observed with other triansa compounds. Cleavage of the ketal with hydrobromic acid in propionic acid, dehydrogenation, hydrolysis, and reductive acetylation, were carried out as described in the synthesis of [2]-catenanes.[16] The reaction sequence led to a mixture of [3]-catenanes **519** and **520** in 32 % yield (based on **515, 518**). Proof for the complete detachment of the double-bridge systems was given by a NMR spectrum of the mixture of **519** and **520**, which showed the expected ratio of proton signals.

13

Investigations of the Synthesis of [3]-Catenanes according to Procedure 2

13.1. Preliminary Investigations

To carry out Procedure 2 on the basis of the reactions worked out in the synthesis of [2]-catenanes, it was planned to synthesize a [3]-precatenane **525** by attaching, via steps **522–524**, twice two additional bridges to a tetrahydroxy-*m*-cyclophane **521**.[185] By using this procedure the formation of isomers, as in the first method, is avoided.

To begin with, it had to be established whether it still was possible to attach twice two bridges to a tetrahydroxy-*m*-cyclophane. To elucidate this basic question, it was attempted to attach twice two bridges to an α,ω-bis-(3,4-dihydroxyphenyl)alkane of the types **526** and **527**. Furthermore, it was investigated whether bisamines **532** and **533** cyclize explicitly to the bisdiansa compounds **534** and **535**, or whether extensive cross cyclization takes place. By cross cyclization it is meant that the amino group does not react with the alkyl halide substrate on the same benzene ring but instead with an alkyl halide substrate attached to the opposite one. Since this cross cyclization can only take place if both parts of the molecule can approach close enough to each other, the length of the chains which join the two benzene rings in compounds of the type **532** and **533** is of utmost importance.

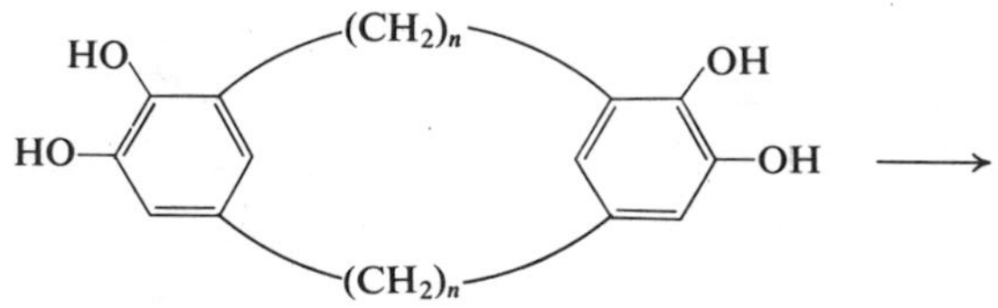

521

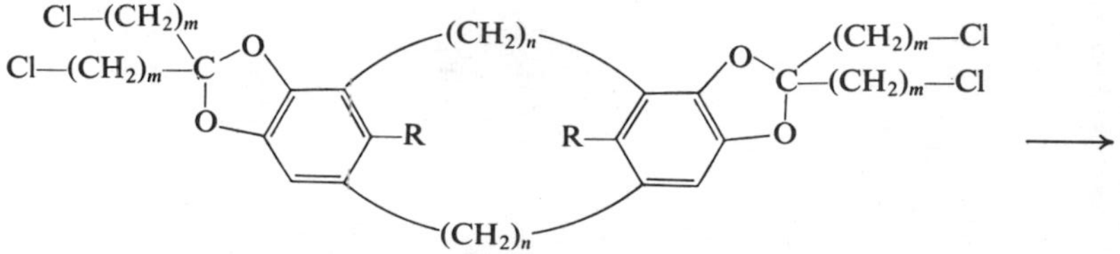

522: R = H
523: R = NO$_2$; $m \geqslant 10$
524: R = NH$_2$; $n \geqslant 10$

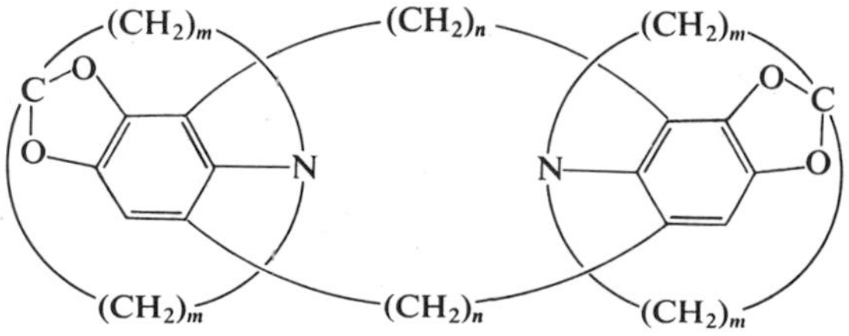

525

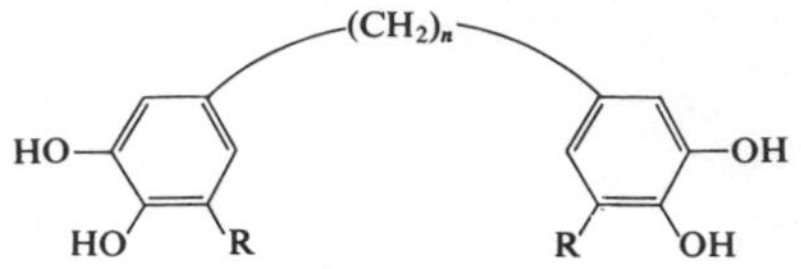

526a, b, c: R = H
527c: R = CH$_3$

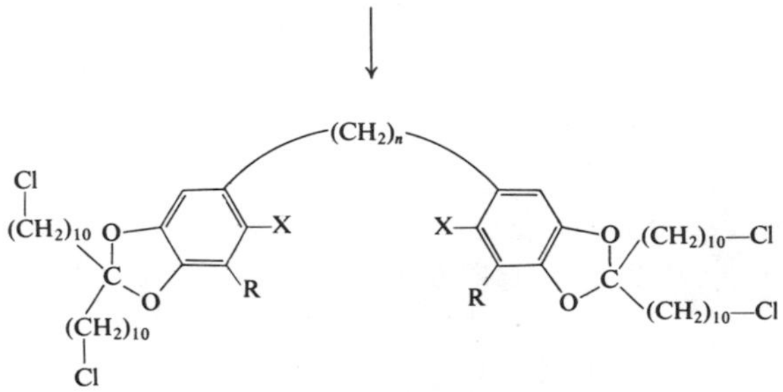

528a, b, c: R = H; X = H
529c: R = CH$_3$; X = H
530a, b, c: R = H; X = NO$_2$
531c: R = CH$_3$; X = NO$_2$
532a, b, c: R = H; X = NH$_2$
533c: R = CH$_3$; X = NH$_2$

a: $n = 2$
b: $n = 6$
c: $n = 12$

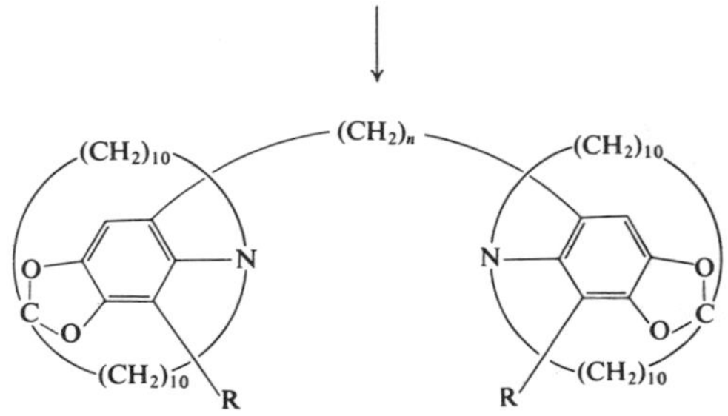

534a, b, c: R = H
535c: R = CH$_3$

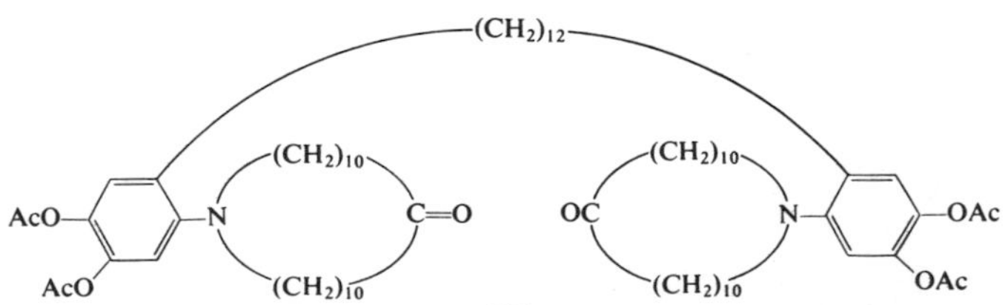

536

Even though one can estimate from molecular models how far the two parts of the molecules can approach each other when joined by a given chain length, it is apparent that the conformation of the paraffin chain under cyclization conditions will play an important part in determining this distance, i.e., the separation will depend upon whether the linking chains are extended or doubled up.

The preliminary investigations were carried out on α,ω-bis(3,4-di-hydroxyphenyl)alkanes **526a, b, c**, and **527c**. The compounds were ketalized according to Salmi[93] with 1,21-dichloroheneicosan-11-one to ketals **528a, b, c**, and **529c**, nitrated with cupric nitrate in acetic anhydride to compounds **530a, b, c**, and **531c** and catalytically reduced with Raney nickel to amines **532a, b, c**, and **533c**. Without previous isolation, the amines were cyclized in isoamyl alcohol under high-dilution conditions in the presence of potassium carbonate[113] and sodium iodide.

In the cyclization of bisamines of the type **532** and **533**, in which the double-bridge systems would contain 10 methylene groups in each bridge, a number of possible isomers can be formed. From the study of molecular models one can classify all the monomeric isomers which could possibly form into three types:

539

1. Alkylation of the amino groups by two alkyl halide substrates from the same aromatic ring results in compounds of the type 534 and 535.

2. Alkylation of the amino groups by alkyl halide substrates from both aromatic rings may result in two isomeric structures 537 and 538. The difference between them is of a similar type as in *cis–trans* isomerism.

3. Alkylation of the amino groups by both alkyl halide substrates of the other aromatic ring may result in compounds of the type 540.

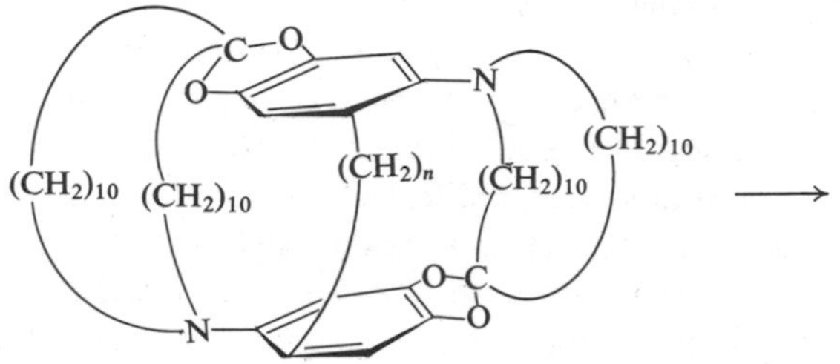

540

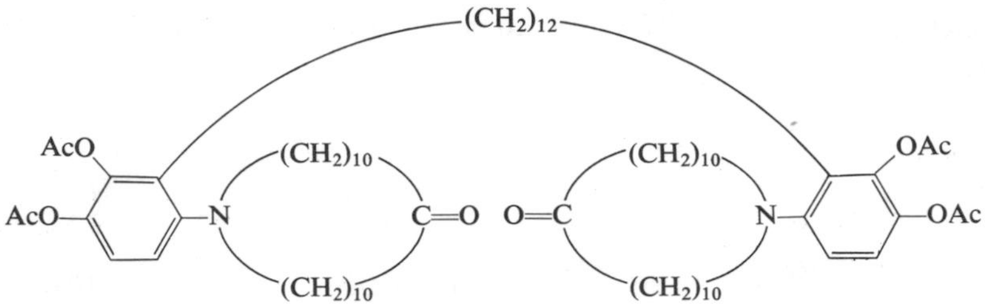

536

The results of the cyclization reactions are summarized in Table 2. Although the yields varied, there was in all cases a preponderance of a distinct reaction product. With diamines 534c and 535c small amounts of a second cyclization product could be isolated. Thin-layer chromatograms of the cyclization products of diamines 532a, b showed, in each case, an additional spot of low intensity which from its R_f value, may have been due to an isomer. Due to the small amounts formed these substances were not isolated.

From the mass spectra, structure interpretations for the amine cyclization products could be carried out. In 534a the cyclization product of diamine 532a, a rupture of the bond between the two methylene groups is especially favored because benzyl cations 541 are thereby formed. Hence, if 534a was the correct structure for the cyclization product, a fragment of half the molecular weight would be predicted. As expected, the spectrum

TABLE 2

CYCLIZATION RESULTS OF SOME DIAMINES

Reacted diamine	Cyclization products	Yield % (based on the corresponding dinitro compounds)
532a	534a	25
532b	534b	7
532c	534c	27
	537 + 538 ($n = 12$)	2
533c	535c	11

showed such a peak; it turned out to be the base peak. The absence of the corresponding isotope peak eliminated the possibility that this peak was due to a double-charged molecular ion.

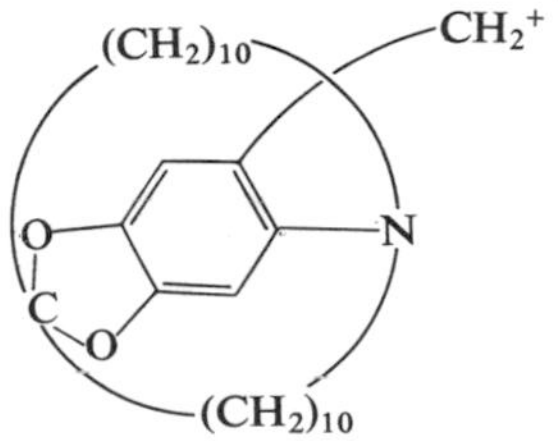

541

The above discussion of compound **534a** also applies to the other cyclization products, with the exception that the fragment ions resulting in benzyl cations of the type **541** are less favored and therefore will appear less intensive in the spectrum. From the analyses of the mass spectra of the other main cyclization products, the respective structures **534b, c**, and **535c** could be deduced. Acid catalyzed hydrolysis of compound **534c** and subsequent acetylation led to compound **536**.

On the basis of the mass spectrum no definite structure could be assigned to the by-product from the cyclization of diamine **532c**. The spectrum showed, however, that it was a monomer. Hydrolysis and acetylation led to a compound which from its infrared spectrum and melting point was isomeric to tetraacetate **536**. Due to the small amount of material available no final decision for either of the two structures could be reached. The cyclization products which are still left to consider are **537**, **538**, and **540** ($n = 12$). On hydrolysis and acetylation the first two will give tetraacetate **539**, while the third will give tetraacetate **536**. For steric reasons, the by-product of the cyclization of diamine **532c** probably has structure **537** or **538** ($n = 12$) or possibly it consists of a mixture of both.

[2]-Rotaxanes

14

Methods
for the
Synthesis
of Rotaxanes

Rotaxanes are mechanically linked aggregates consisting of a macrocycle and a threaded chain with bulky end groups. They can be isolated if the energy barrier, which has to be surmounted in order to separate the subunits, reaches a certain minimal value. This requirement generally means that a certain relation exists between the space requirements of the two subunits. When a cycloparaffin and a paraffin chain are used, the macrocycle must, as in the case of a catenane, consist of at least 18 methylene groups in order that the paraffin chain may be threaded through it. Similarly to catenanes, rotaxanes can be synthesized by statistical as well as directed methods.[25]

14.1. Statistical Methods

When the statistical method is carried out with the proper starting materials, a rotaxane is formed alongside its molecular subunits in a ratio dependent on statistical principles. Proposals for the statistical synthesis of rotaxanes can, as in the case of catenanes, be divided into various schemes.

14.1.1. *Scheme 1*

In presence of macrocycle **37**, a chain **36** is bonded to the bulky end groups **542**. Thereby a certain amount of rotaxane **2** may form alongside

the unchanged macrocycle **37** and the dumbbell-shaped molecule **543**. Similar results should be obtained on cyclizing a chain **36** in a solvent of dumbbell-like molecules **543**.

14.1.2. *Scheme 2*

An increase in the yield of rotaxane **2** can be achieved by temporarily linking the chain **36** to the macrocycle **37**. The extra- and intraannular isomers **544** and **545** thus obtained, would exist in a conformational equilibrium. After attaching bulky end groups **542** and cleaving the temporary bonds, a mixture of rotaxane **2**, compound **543** and macrocycle **37** is obtained.

14.1.3. *Scheme 3*

A dumbbell-shaped compound **543** is linked to the bifunctional chain **36** to give **546**. When the groups X react with each other, a mixture of the intra-

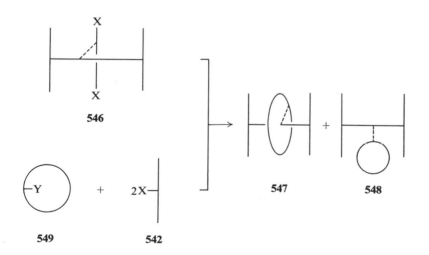

and extraannular compounds **547** and **548** is obtained. After cleaving the temporary linkages, a mixture of rotaxane **2** and compounds **543** and **37** is obtained. Similar results are obtained by attaching two bulky groups **542** to a macrocycle **549** with substituent Y.

14.2. Directed Synthesis

In the directed synthesis of rotaxanes, as was already mentioned in Section 7.2, a sterically fixed intraannular arrangement between a macro-cycle and a chain is brought about, by taking advantage of the tetrahedral structure of the carbon atoms in combination with certain rigid structures. A rotaxane is obtained by the attachment of bulky end groups to the chain and subsequent cleavage of the linkages between the aromatic nucleus and the macrocycle.

15

Investigations of the Statistical Syntheses of Rotaxanes

15.1. Scheme 1

This scheme for the synthesis of rotaxanes has been investigated cursorily by Freudenberg and Cramer,[27] as well as by Stetter and Lihotzky,[28] and more in depth by Harrison and Harrison.[29] Harrison and Harrison[29] reacted 2-hydroxycyclotriacontanone **550a** with succinic anhydride and thus obtained the half ester **550b**. The sodium salt of this compound was converted to the resin bond macrocycle **550c** by reacting it with the chloromethylated copolymer from styrene and divinylbenzene. The latter was obtained by the method of Merrifield.[198] The resin bound macrocycle was treated 70 times with 1,10-decanediol and triphenylmethyl chloride in a mixture of pyridine, dimethylformamide, and toluene. Unfortunately, the amounts of reaction materials used were not mentioned in the communication. After hydrolysis with sodium bicarbonate in refluxing methanol, a mixture containing 6% of rotaxane **550d** was obtained. Following chromatography, the rotaxane was isolated as an oil stable up to 200°. The compound showed the expected infrared absorption and when subjected to thin-layer chromatography was shown to be free of unbound molecular subunits.

The structure of rotaxane **550d** was verified by chemical degradation. Oxidation with silver oxide gave octacosane-1,28-dicarboxylic acid, which was isolated as the dimethyl ester and the dumbbell-shaped molecule **550f**. Cleavage of rotaxane **550d** with BF_3-etherate in benzene resulted in decane-1,10-diol, triphenylmethanol, and acyloin **550a**.

550a: R = H
550b: R = CO—$(CH_2)_2$—COOH
550c: R = CO—$(CH_2)_2$—COO—resin

$$(H_5C_6)_3C—O——(CH_2)_{10}——O—C(C_6H_5)_3$$
550f

Freudenberg and Cramer[27] attempted to react inclusion compounds of cyclodextrins and long-chain dialdehydes with bulky hydrazine derivatives such as naphthylhydrazine. By inspecting a Stuart-Briegleb molecular model it can be seen, however, that naphthalin derivatives are not bulky enough to prevent the chain from slipping out of the cyclodextrin ring. If for no other reasons than this, the synthesis could not be successful.

Stetter and Swincicki[199] attempted to synthesize a rotaxane by reacting hexamethylene-1,6-diamine with diphenylketene in the presence of macrocycle **551**. The reactions were carried out without solvent and in the presence of tetralin; here again, no rotaxane could be isolated.

551

The investigations of Stetter and Lihotzky[28] were based on the observation that bis(N,N-tetramethylenebenzidine) **552** forms adducts with dioxane and benzene.[138] In further investigations it was found that the tetraformyl compound **553** crystallizes with a molecule of solvent from pyridine. When compound **553** as well as the tetraacetyl compound **554** were dissolved in molten benzidine and washed free of excess solvent, they were found to retain a molecule of benzidine. By heating the adduct in a vacuum at 200° the benzidine could be recovered. The benzidine is also displaced by dissolving the adduct in diethylene glycol; now a compound is obtained containing four molecules of solvent.

In order to obtain a rotaxane, the 1:1 adducts of **553** and **554** with benzidine were reacted with triphenylpropyl isocyanate **555** in a molten state, as well as in the solvents benzene, dioxane, and anisole. Benzidine was also reacted with compound **555** in anisole in the presence of **554**. In all cases only the unchanged macrocycle **554** and the diurea compound **556** could be isolated. Likewise, when compound **554** was present in the reaction of hexamethylene-1,6-diamine with **555**, or in the reaction of terephthalyl dichloride with triphenylpropylamine, only compounds **557** and **558** were

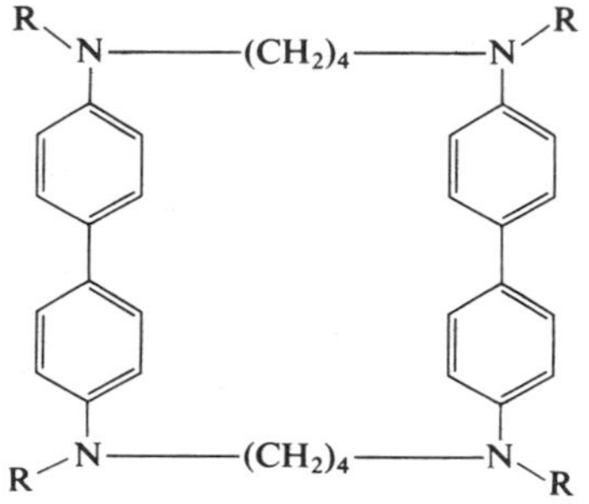

552: R = H
553: R = CHO
554: R = COCH$_3$

$(C_6H_5)_3C—CH_2—CH_2—N=C=O$
555

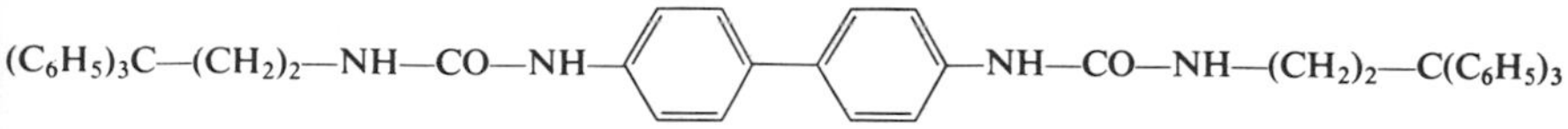

$(C_6H_5)_3C—(CH_2)_2—NH—CO—NH$⟨⟩⟨⟩$—NH—CO—NH—(CH_2)_2—C(C_6H_5)_3$
556

$(C_6H_5)_3C—(CH_2)_2—NH—CO—NH—(CH_2)_6—NH—CO—NH—(CH_2)_2—C(C_6H_5)_3$
557

$(C_6H_5)_3C—(CH_2)_2—NH—CO$⟨⟩$—CO—NH—(CH_2)_2—C(C_6H_5)_3$
558

$(C_6H_5)_3C—(CH_2)_2—NH—CO—NH—(CH_2)_2—C(C_6H_5)_3$
559

obtained, respectively. Similarly, reaction of **555** and triphenylpropylamine in the presence of **554** only resulted in the urea derivative **559**.

15.2.   Scheme 2

Studies which can be schematically classified into this group were
carried out by Schill and Tafelmair.[126] Based on the work discussed in
Section 8.5, 2,5-dibromohydroquinone was reacted with excess 1,10-
dibromodecane to diether **560**. Alkylation of 3,4,5-triphenylphenol with
this compound resulted in the dumbbell-shaped molecule **561**. Treatment
with cuprous cyanide in dimethylformamide according to Friedman and
Shechter[125] gave dinitrile **563**. In a second procedure this compound was
prepared by reacting 2,5-dicyanohydroquinone with excess 1,10-dibromo-
decane and subsequently treating compound **562** thus obtained, with 3,4,5-
triphenylphenol. Dinitrile **563** was reduced with LiAlH$_4$ to diamine **564**
and then reacted with 11-chloroundecanoyl chloride to give diamide **565**.
This compound could be reduced to diamine **566** with diisobutyl aluminum
hydride, or better yet, with diborane in tetrahydrofuran. Diamine **566** was

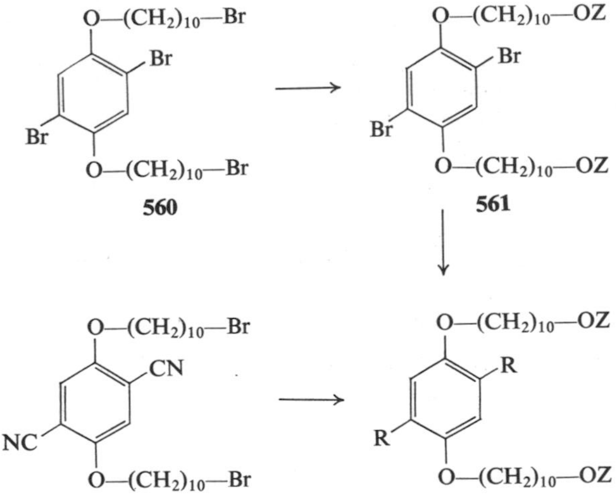

560

561

562

563: R = CN
564: R = CH$_2$—NH$_2$
565: R = CH$_2$—NH—CO—(CH$_2$)$_{10}$—Cl
566: R = CH$_2$—NH—(CH$_2$)$_{11}$—Cl

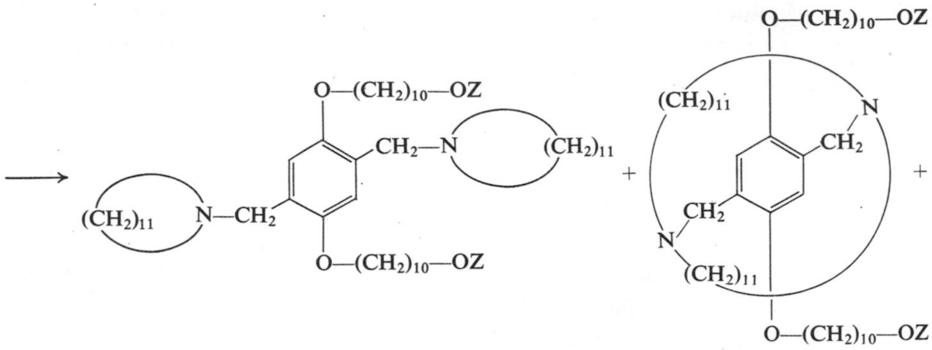

567   568

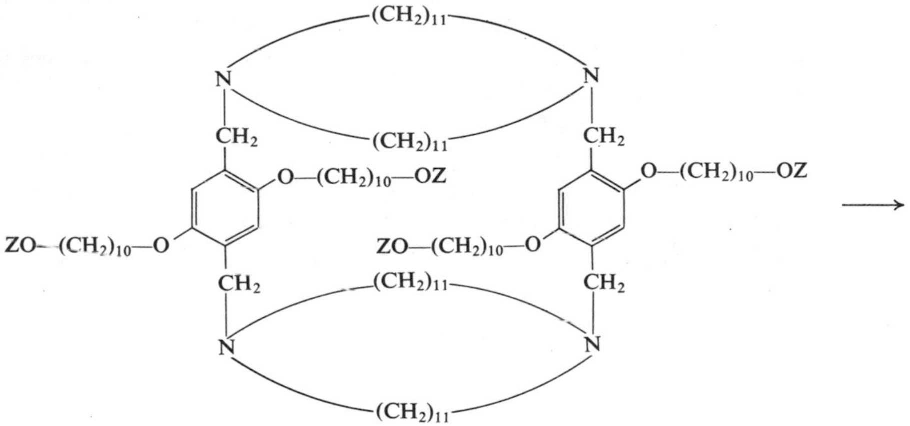

569

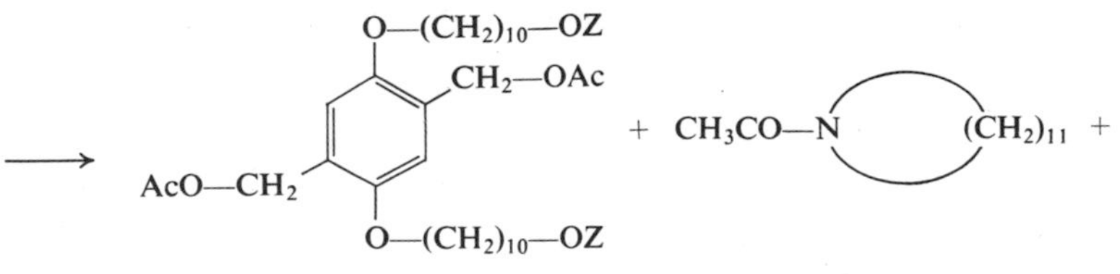

570   193

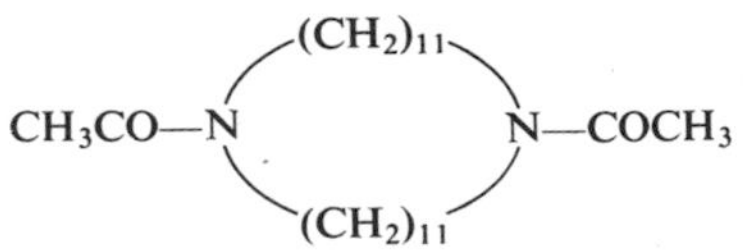

204

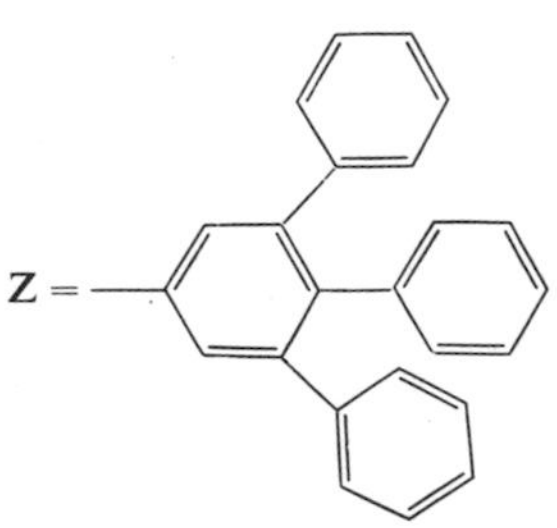

$$Z = $$

cyclized in isoamyl alcohol by the high-dilution method in the presence of potassium carbonate and sodium iodide. Acetylation with acetic anhydride in pyridine and chromatography gave, in 21% yield, a mixture of three amines. From the chromatographic behavior, inspection of the mass spectra, chemical degradation by acetolysis, and comparison with independently synthesized compound 567, structures 567, 568, and 569 were assigned to the three products. Since the separation of the mixture could not be carried out on a preparative scale, the cyclization products were cleaved with acetic anhydride. Compounds 570, 193, and 204 could be isolated by thin-layer chromatography. A rotaxane could not be detected.

Of interest is the structure of the dimeric product 569. It differs from the dimer 206, obtained from the model compound 202 (see Section 8.5), in that both aromatic rings are now joined by four chains. The formation of this isomeric structure probably is caused by the bulky phenol substrates, which favor the formation of a structure in which the hydroquinone nuclei are separated as far as possible.

To obtain compound 567, required for comparison, dinitrile 563 was hydrolyzed to the dicarboxylic acid 571 and subsequently esterified to 572. Reduction resulted in 573 and acetylation gave compound 574. For the preparation of the cyclization product 567, diol 573 was converted to dibromide 575 with triphenylphosphine dibromide[193] and then reacted with cycloundecylamine.

$$O-(CH_2)_{10}-OZ$$

$$R$$

$$R$$

$$O-(CH_2)_{10}-OZ$$

571: $R = COOH$
572: $R = COOC_2H_5$
573: $R = CH_2-OH$
574: $R = CH_2-O-COCH_3$
575: $R = CH_2-Br$

16

Directed
Synthesis
of
Rotaxanes

Investigations of the directed synthesis of rotaxanes were carried out by Schill and Zollenkopf.[25, 26] Bifunctional diansa compound **449**, whose synthesis was already described in Section 9.6.2, was used to alkylate the sodium salt of *N*-acetyl-2,4,6-tri-*p*-tolylaniline in dimethylformamide to give compound **576** in 81 % yield. Confirmation of the assumed structure was obtained by mass spectrometry. It is remarkable that the spectrum showed the molecular ion peak at $m/e = 1611$ with the relative high intensity of 27 % of the base peak.

The conversion of prerotaxane **576** to a rotaxane was carried out in analogy to the synthesis of catenanes.[16] Ketal cleavage with hydrobromic acid in acetic acid gave aminocatechol **577** and partially deacetylated products. The reaction mixture was then dehydrogenated with ferric sulfate to the amino-*o*-benzoquinone. Through acid hydrolysis the corresponding hydroxy-*o*-benzoquinone is formed; it spontaneously tautomerizes to the *p*-quinoid structure **578**. Besides the compounds **577** and **578**, the reaction mixture also contained the partially deacetylated derivatives. Acetylation of the reaction mixture resulted in tetraacetate **579** from which the hexaacetate **580** was prepared by reductive acetylation.

The following arguments in favor of the assumed rotaxane structure **580** were brought forward:

1. The elemental analysis is satisfactory.

2. The IR spectrum shows the expected bands. The spectra of the rotaxane and the equimolar mixture of the two single components are identical except for one weak additional band in the mixture.

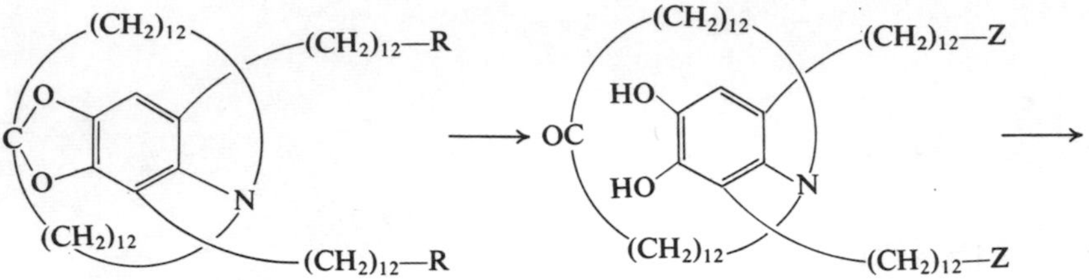

449: R = Br
576: R = Z

577

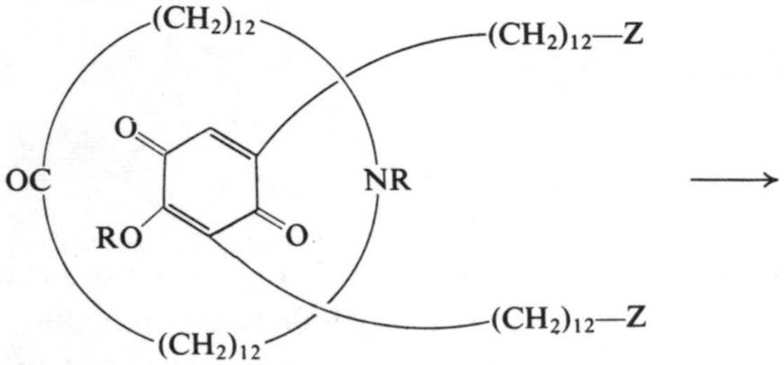

578: R = H
579: R = COCH$_3$

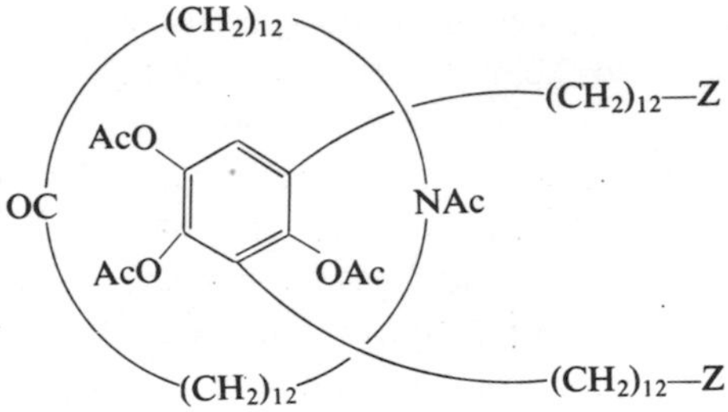

580

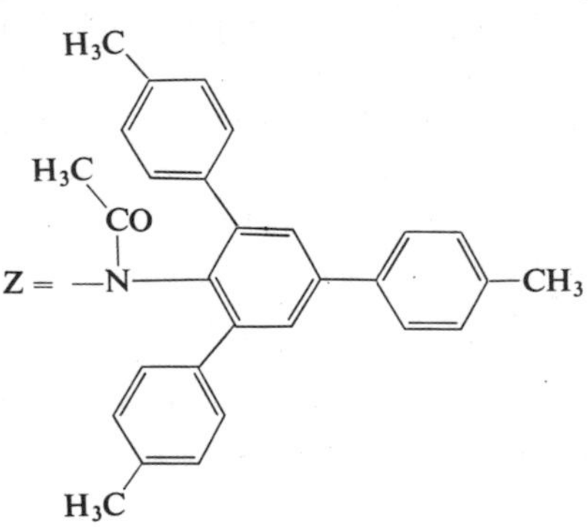

3. In thin-layer chromatograms the rotaxane is homogenous and has a R_f value clearly different from the R_f values of its individual components.

16.1. Synthesis of Molecular Subunits of Rotaxanes

The synthesis of the macroheterocycle **390b** has already been discussed in Sections 9.4.2 and 9.8. For the synthesis of the dumbbell-shaped com-

581: R = R′ = H
582: R = COCH$_3$; R′ = H
583: R = COCH$_3$; R′ = NO$_2$
584: R = H; R′ = NO$_2$
585: R = H; R′ = NH$_2$

586: R = OH; R′ = H
587: R = OH; R′ = n-C$_4$H$_9$
588: R = OCOCH$_3$; R′ = n-C$_4$H$_9$
589: R = OCOCH$_3$; R′ = COCH$_3$
590: R = Br; R′ = n-C$_4$H$_9$
591: R = Z; R′ = n-C$_4$H$_9$

592: R = H; R′ = N(n-C$_4$H$_9$)$_2$
593: R = COCH$_3$; R′ = OCOCH$_3$

ponent **593**, catecholdiol **433** (see Section 9.5) was ketalized with diethyl ketone to **581**, acetylated to **582**, and nitrated in the 5-position with cupric nitrate in acetic anhydride[109] to **583**. Hydrolysis led to diol **584** which was converted to amine **585** by catalytic reduction. The alkylation with butyl bromide was carried out in refluxing isoamyl alcohol in the presence of potassium carbonate and sodium iodide. Even though an excess of alkyl bromide was used, only a mixture of mono- and dialkylamino compounds **586** and **587** was obtained. The former predominated in this mixture. After acetylation, the tertiary amine **588** could be separated from amide **589**. Saponification of **588** led to diol **587** which was converted with triphenyl-phosphine dibromide[193] to dibromide **590** in 25% yield. Reaction with the bulky end groups gave **591**, subsequent ketal cleavage led to the catechol derivative **592** which was converted, in analogy to the rotaxane, to penta-acetate **593**.

Knots, Double Wound Catenanes and Rotaxanes, and Higher Linear Catenanes

17

Concepts
of
Syntheses

17.1. Concept for the Synthesis of a Knot

On the basis of statistical laws, a bifunctional chain of suitable length, when cyclized, should form small amounts of a trefoil knot **3**.[3] Judging from a molecular model, at least about 50 methylene groups are necessary for the synthesis of such a knot. A general method for the synthesis of a trefoil knot consists in transferring the principles of a Möbius strip to a chemical system. The theoretical possibilities have been discussed in detail especially by van Gulick.[72] No experimental studies concerned with these theories or suggestions which are chemically feasible have been reported as yet.

Schill and co-workers[30, 31] designed a concept for the synthesis of a trefoil knot and investigated with model reactions an experimental sequence which might lead to such a structure. The reaction path was designed such that a knot, a double wound catenane, and higher linear catenanes might all be synthesized from a common precursor. A suitable compound for such a precursor was found in the intraannularly bonded diansa compound **594**. Cyclization of one functional group X with the more distant amino group in diansa compound **594** leads to compound **595**. A second cyclization results in the topological isomers **596** and **597**. Cleavage of the ketal linkage as well as the aryl–nitrogen bonds gives a trefoil knot **598** and a simple macrocycle **599**.

594

595

596

597

598

599

17.2. Concept for the Synthesis of Double Wound Catenanes and Rotaxanes

A. A theoretically possible synthesis of double wound catenanes and rotaxanes begins with the cyclization of functional groups X in diansa compound **594**. The attachment of two alkyl substrates to the resulting

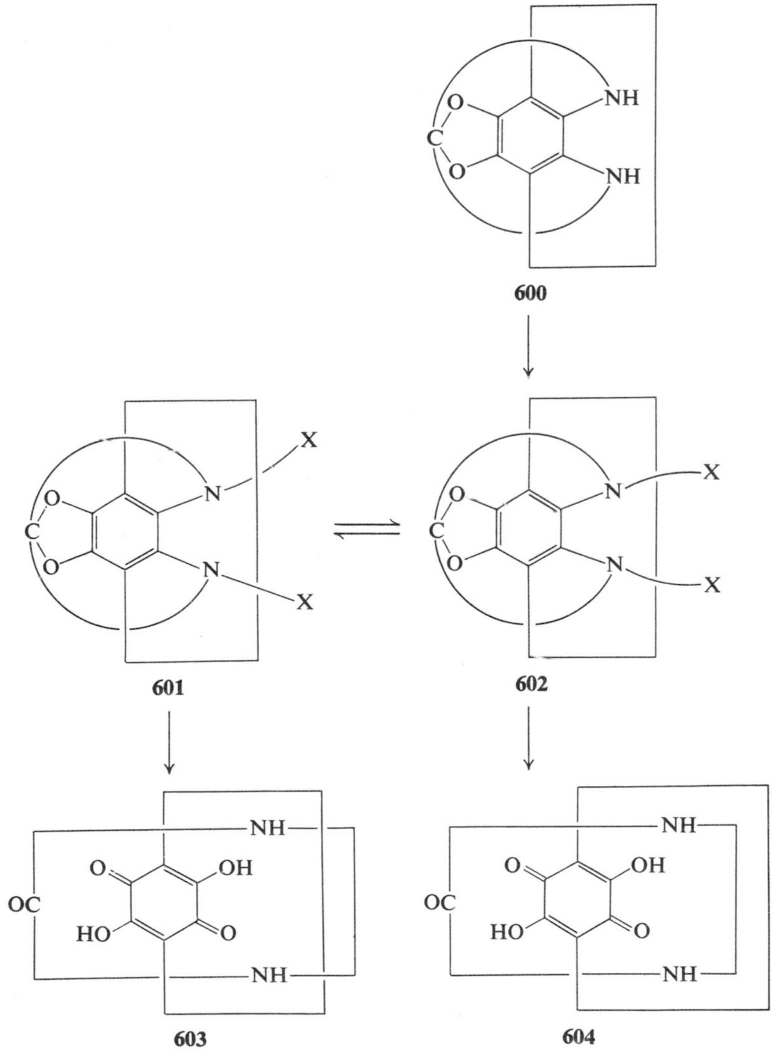

600

601

602

603

604

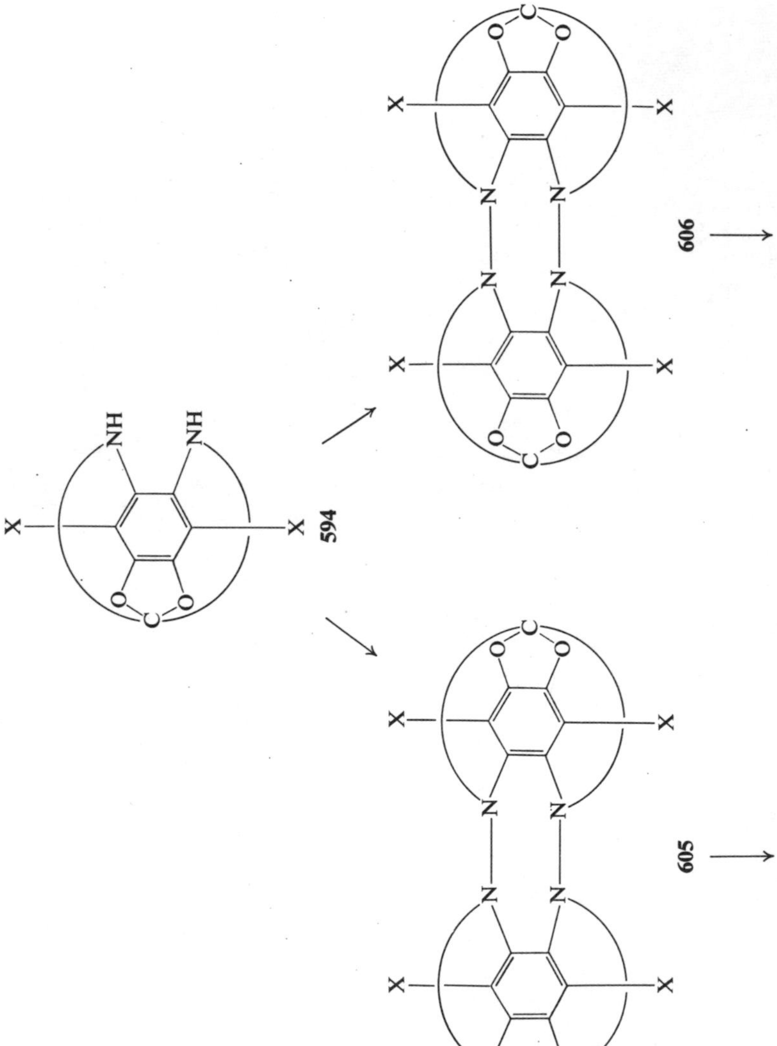
594
605
606

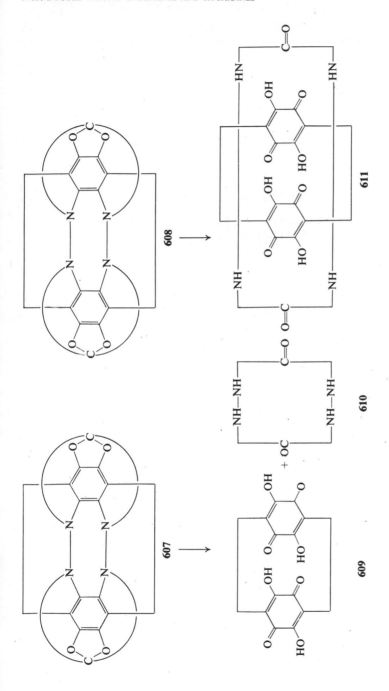

triansa compound **600** gives a new compound, two conformers of which are represented by formulas **601** and **602**. The equilibrium is probably shifted toward compound **602**. After cyclization of the functional groups X and cleavage of the ketal and aryl–nitrogen bonds, simple catenane **604** or double wound catenane **603** is obtained, depending on the conformation in which the cyclization takes place. In a similar manner, a double wound rotaxane is obtained if functions X represent bulky end groups.

B. The linkage of two diansa compounds **594** by the two nitrogen atoms via two chains, gives stereomers **605** and **606**. Compound **606** is a racemate and **605** a *meso* form. One optical isomer of compound **606** is obtained as sole reaction product when two molecules of one enantiomer of compound **594** are linked together. When the outer bridges of compounds **605** and **606** are closed, topological isomers **607** and **608** are formed, and if the ketal and aryl–nitrogen bonds are cleaved, catenane **611** with winding number $\alpha = 2$, and macrocycles **609** and **610** are obtained. Similarly, a simple rotaxane and a double wound rotaxane can be obtained according to this scheme.

17.3. Concept for the Synthesis of Higher Linear Catenanes

The principle for the directed synthesis of catenanes found by Schill[69] is applicable to the synthesis of two, three, and higher clustered catenanes as for example **612**, but it can not be used for the synthesis of a continuous linear chain of linked macrocycles. Such a synthesis can be designed if diansa compound **594** is used as starting material. The intermolecular cyclization of diansa compound **594** leads to compound **613**. Continued repetition of these cyclizations results first in compound **614** and then in oligomeric linear precatenanes. Cyclization of the functional end groups X

612

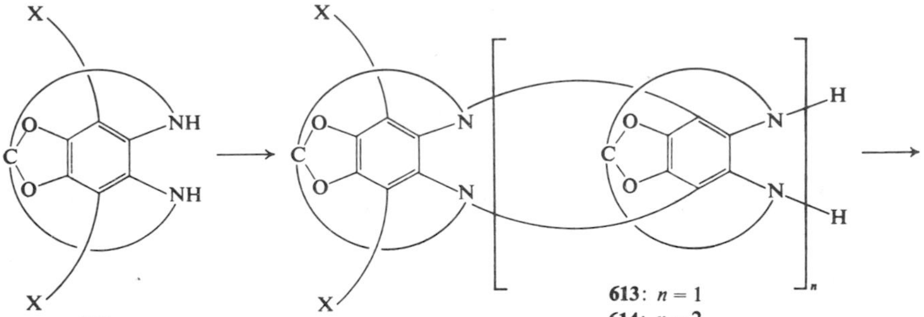

613: $n = 1$
614: $n = 2$

594

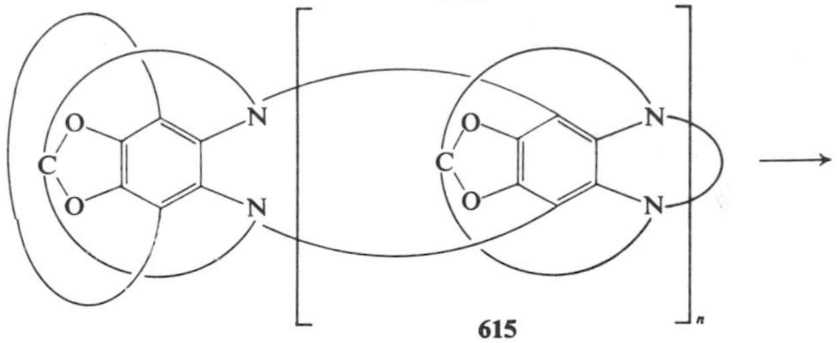

615

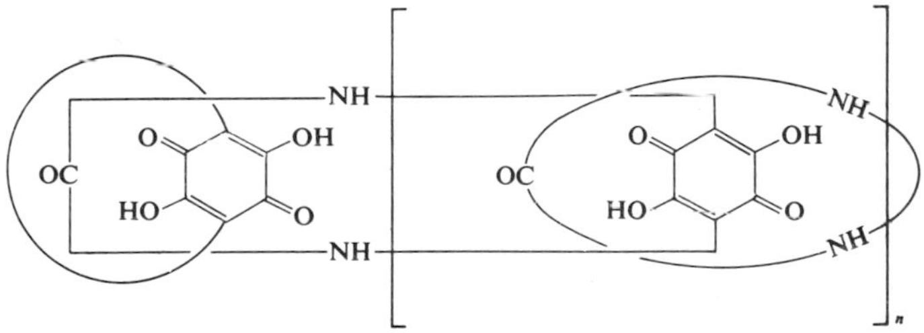

616

and closure of the last diansa bridge by joining the two amino groups with a chain results in a compound of the type **615**. Cleavage of the ketal and aryl–nitrogen linkages yields oligocatenane **616**. Intramolecular cyclization of compounds of the type **614** ($n \geqslant 2$) results in cyclocatenanes.

18

Investigations of the Synthesis of Knots, Double Wound Catenanes and Rotaxanes, and Higher Linear Catenanes

18.1. Model Investigations

The key substance for the above-mentioned synthesis is diansa compound **594**. Model investigations of the synthesis of such a substance were carried out by Schill, Henschel, and Boeckmann.[30] Ketalization of 3,6-dimethyl catechol with 1,21-dichloroheneicosan-11-one yields ketal **617**, which can be converted to the dinitro compound **618** with fuming nitric acid. Catalytic reduction with Raney nickel or platinum led to diamine **619** and subsequent reaction with tosylchloride to bistosylamide **620**. Diansa compound **622** is obtained in 30% yield (based on **620**) by cyclizing diiodide **621** using the high-dilution method in dimethyl sulfoxide in the presence of potassium carbonate. Detosylation with sodium in butanol[135] gives diamine **623**.

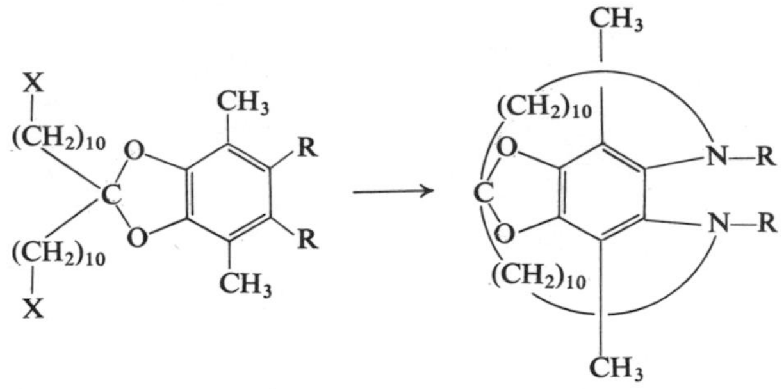

617: R = H; X = Cl
618: R = NO$_2$; X = Cl
619: R = NH$_2$; X = Cl
620: R = NH—Tosyl; X = Cl
621: R = NH—Tosyl; X = I

622: R = Tosyl
623: R = H

19

Tables
of
Compounds

In the following tables only isolated compounds have been listed.

DI- AND TRIANSA COMPOUNDS OF THE TYPE

R	R'	n	References
H	H	10	113
H	OH	10	38, 131, 146
H	$OCOCH_3$	10	146
H	OCH_3	10	38, 131
H	OC_2H_5	10	69
H	$O-(CH_2)_{11}-OH$	10	69
H	$(CH_2)_8-COOC_2H_5$	10	38
H	$(CH_2)_{10}-COOC_2H_5$	10	38
H	$(CH_2)_{10}-CN$	10	38, 131

	R'——R'		
H	$O-(CH_2)_{21}-O$	10	38, 69, 131
H	$O-(CH_2)_{10}-\underset{\underset{NH_2}{\mid}}{C}=\underset{\underset{CN}{\mid}}{C}-(CH_2)_9-O$	10	38, 131
H	$O-(CH_2)_{10}-\overset{\overset{O}{\mid\mid}}{C}-(CH_2)_{10}-O$	10	38, 131

R	R'		
CH_3	OCH_3	11	159
CH_3	OH	11	159
CH_3	OCH_3	13	159
CH_3	OH	13	159
$(CH_2)_3-OH$	OCH_3	13	159

DIANSA COMPOUNDS OF THE TYPE

R	References
Br	38
OCH$_3$	38, 131

DI- AND TRIANSA COMPOUNDS OF THE TYPE

R	n	References
H	10	165
CH_3	10	165
C_3H_7	10	165
C_3H_7	12	16
$(CH_2)_{12}$—OH	10	25
$(CH_2)_{12}$—OH	12	25, 188
$(CH_2)_{12}$—OCOCH$_3$	10	25
$(CH_2)_{12}$—OCOCH$_3$	12	25, 188
$(CH_2)_{12}$—Cl	10	25
$(CH_2)_{12}$—Cl	12	25
$(CH_2)_{12}$—Br	12	25, 188

$(CH_2)_{12}$—N—Z^a
       |
     COCH$_3$
	12	25

R———R	n	References
$(CH_2)_{25}$	10	16
$(CH_2)_{25}$	12	16, 18
$(CH_2)_{12}$—NH—$(CH_2)_{12}$	12	188

COCH$_3$
      |
 $(CH_2)_{12}$—N—$(CH_2)_{12}$
     Tosyl
	12	188

| $(CH_2)_{12}$—N—$(CH_2)_{12}$	12	188

^{a} Z = 2, 4, 6-tri-p-tolyl

DIANSA COMPOUNDS OF THE TYPE

R	References
H	30
Tos	30

MISCELLANEOUS DIANSA COMPOUNDS

Structure	n	R	References

	10	H	146
	11	CH_3	157
	13	CH_3	157

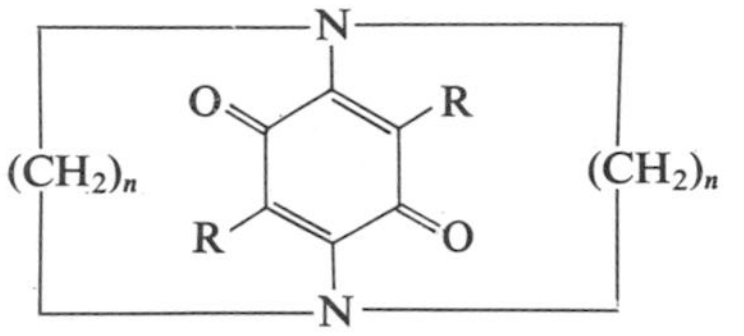

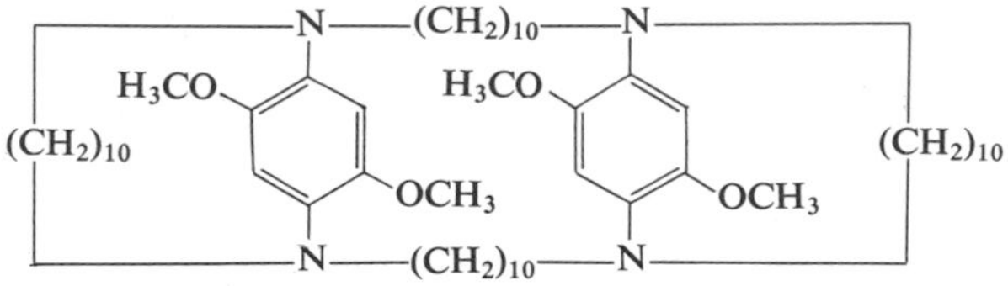

38

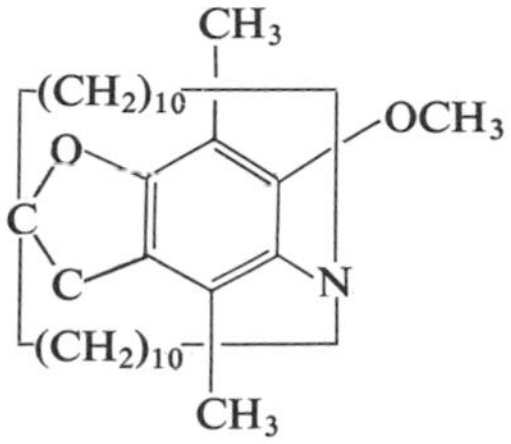

166

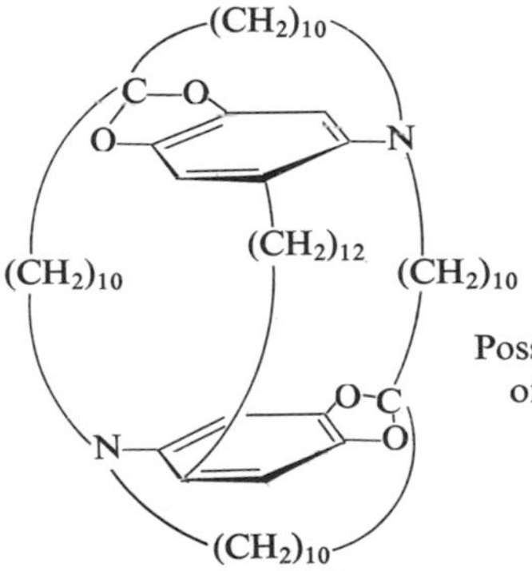

Possibly mixture
of stereoisomers

185

BISDIANSA COMPOUNDS OF THE TYPE

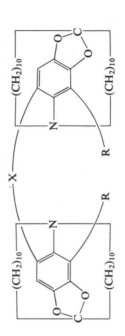

R	X	n	References
H	$(CH_2)_2$	10	185
H	$(CH_2)_6$	10	185
H	$(CH_2)_{12}$	10	185
CH_3	$(CH_2)_{12}$	10	185
$(CH_2)_{12}$—NH—$(CH_2)_{12}$ COCH_3	$(CH_2)_{12}$—NH—$(CH_2)_{12}$ COCH_3	12	
—N—$(CH_2)_{12}$ Tosyl	—N—$(CH_2)_{12}$ Tosyl	12	188, 197
—N—$(CH_2)_{12}$	—N—$(CH_2)_{12}$	12	

Mixture of structural isomers

R——R

INTRAANNULARLY BONDED RING SYSTEMS OF THE TYPE

R	n	References
H	12	15, 16
COCH$_3$	12	15, 16
H	16	20, 21
CH$_3$	16	20, 21

[2]-CATENANES OF THE TYPE

X	n	References
CH$_2$	12	15–19
N—COCH$_3$	12	188
CH$_2$	17	20, 21

[2]-CATENANES OF THE TYPE

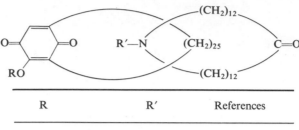

R	R′	References
H	COCH₃	15, 16, 18
COCH₃	COCH₃	15, 16, 18

[3]-CATENANES

Structure	References
Mixture	188, 197

MISCELLANEOUS CATENANES

Structure	References

$C_{34}H_{63}D_5$ C=O HC—OH (CH$_2$)$_{32}$ 1, 14, 74

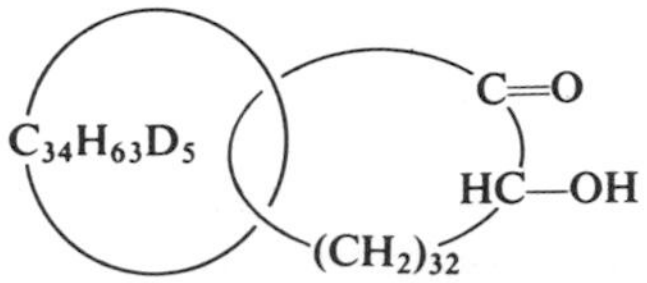

[2]-[cyclic 186 phage DNA]-   22
    [cyclic coliphage λ DNA]-catenane
Naturally occurring catenanes in
    mitochondrial DNA from
        HeLa cells   23
        Leukemic leukocytes   24
        3T3 Mouse cells transformed by SV40 Virus   24, 50b
        Unfertilized eggs of sea urchins   24, 50b
    in kinetoplastic DNA from *Trypanosoma cruzi*   51

[2]-ROTAXANES

Structure	References

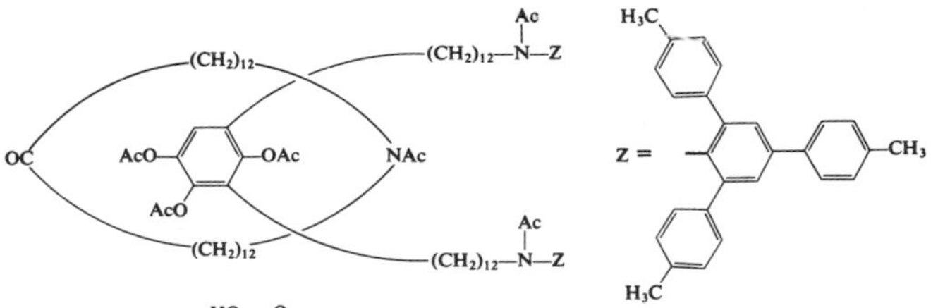

25

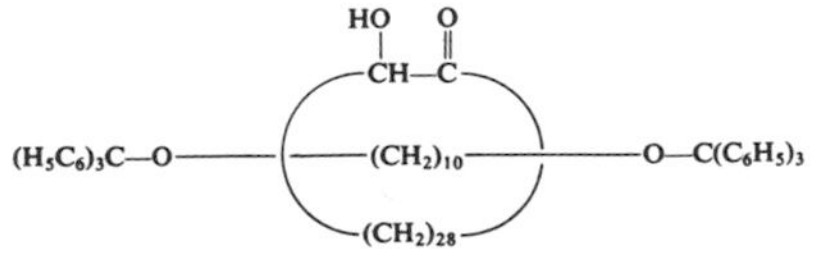

29

BIBLIOGRAPHY

1. E. Wasserman, *J. Amer. Chem. Soc.* **82**, 4433 (1960).
2. See footnote 5 in Frisch and Wasserman.[3]
3. H. L. Frisch and E. Wasserman, *J. Amer. Chem. Soc.* **83**, 3789 (1961).
4. S. D. Ross, E. R. Coburn, W. A. Leach, and W. B. Robinson, *J. Polymer Sci.* **13**, 406 (1954); H. Zahn, *Angew. Chem.* **68**, 164 (1956); H. Zahn, P. Rathgeber, E. Rexroth, R. Krzikalla, W. Lauer, P. Miró, H. Spoor, F. Schmidt, B. Seidel, and D. Hildebrand, *ibid.* p. 229; H. Zahn, *ibid.* **69**, 270 (1957); C. J. Brown, A. Hill, and P. V. Youle, *Nature (London)* **177**, 128 (1956); W. Patnode and D. F. Wilcock, *J. Amer. Chem. Soc.* **68**, 358 (1946).
5. A. Lüttringhaus, F. Cramer, H. Prinzbach, and F. M. Henglein, *Justus Liebigs Ann. Chem.* **613**, 185 (1958); also see A. Lüttringhaus, F. Cramer, and H. Prinzbach, *Angew. Chem.* **60**, 137 (1957).
6. H. Frisch, I. Martin, and H. Mark, *Monatsh. Chem.* **84**, 250 (1953); H. Mark, *ibid.* **83**, 545 (1952).
7. F. Patat and P. Derst, *Angew. Chem.* **71**, 105 (1959).
8. K. Ziegler, H. Eberle, and H. Ohlinger, *Justus Liebigs Ann. Chem.* **504**, 94 (1933); K. Ziegler and A. Lüttringhaus, *ibid.* **511**, 1 (1934); K. Ziegler and K. Weber, *ibid.* **512**, 164 (1934); K. Ziegler and R. Aurnhammer, *ibid.* **513**, 43 (1934).
9. K. Ziegler and W. Hechelhammer, *Justus Liebigs Ann. Chem.* **528**, 114 (1937).
10. V. L. Hansley, *J. Amer. Chem. Soc.* **57**, 2303 (1935).
11. V. L. Hansley, U.S. Patent 2,228,268 (1941).
12. V. Prelog, L. Frenkiel, M. Kobelt, and P. Barman, *Helv. Chim. Acta* **30**, 1741 (1947).
13. M. Stoll and A. Rouvé, *Helv. Chim. Acta* **30**, 1822 (1947); M. Stoll and J. Hulstkamp, *ibid.* p. 1815.
14. E. Wasserman, *Sci. Amer.* **207**, No. 5, 94 (1962).
15. G. Schill, Habilitation Thesis, University of Freiburg, 1964.
16. G. Schill, *Chem. Ber.* **100**, 2021 (1967).
17. Also see A. Lüttringhaus, *in* "Stereochemie der Kohlenstoffverbindungen" (E. L. Eliel, ed.), p. 252. Verlag Chemie, Weinheim, 1966.
18. G. Schill and A. Lüttringhaus, *Angew. Chem.* **76**, 567 (1964); *Angew. Chem. Int. Ed. Engl.* **3**, 546 (1964).
19. W. Vetter and G. Schill, *Tetrahedron* **23**, 3079 (1967).
20. A. Lüttringhaus and G. Isele, *Angew. Chem.* **79**, 945 (1967); *Angew. Chem. Int. Ed. Engl.* **6**, 956 (1967).
21. G. Isele, Dissertation, University of Freiburg, 1968.
22. J. C. Wang and H. Schwartz, *Biopolymers* **5**, 953 (1967).
23. B. Hudson and J. Vinograd, *Nature (London)* **216**, 647 (1967).
24. D. A. Clayton and J. Vinograd, *Nature (London)* **216**, 652 (1967).
25. G. Schill and H. Zollenkopf, *Justus Liebigs Ann. Chem.* **721**, 53 (1969); also see *Nachr. Chem. Tech.* **15**, 149 (1967).

26. H. Zollenkopf, Dissertation, University of Freiburg, 1968.
27. K. Freudenberg and F. Cramer, see footnote 14 in Lüttringhaus, Cramer, Prinzbach, and Henglein.[5]
28. H. Stetter and R. Lihotzky, unpublished data (1962); also see R. Lihotzky, Dissertation, Technische Hochschule, Aachen, 1962.
29. I. T. Harrison and S. Harrison, *J. Amer. Chem. Soc.* **89**, 5723 (1967).
30. G. Schill, R. Henschel, and J. Boeckmann, to be published.
31. G. Schill, *Int. Symp. Conformational Anal., 1969* in press.
32. R. Henschel, part of the not yet finished Dissertation, University of Freiburg.
33. J. Boeckmann, part of the not yet finished Dissertation, University of Freiburg.
34. A. Lüttringhaus, *in* "Stereochemie der Kohlenstoffverbindungen" (E. L. Eliel, ed.), p. 250. Verlag Chemie, Weinheim, 1966.
35. G. W. Wheland, "Advanced Organic Chemistry," 2nd ed., p. 32. Wiley, New York, 1949.
36. J. Tauber, *J. Res. Nat. Bur. Stand., Sect. A*, **67**, 591 (1963).
37. H. Kohler and D. Dieterich, Deutsche Auslegeschrift Appl. 1,069,617 (1957); also see *Nachr. Chem. Tech.* **8**, 87 (1960).
38. T. Doornbos and J. Strating, unpublished data (1966); T. Doornbos, Ph.D. Thesis, University of Groningen, Netherlands, 1966.
39. E. L. Eliel, ed., "Stereochemie der Kohlenstoffverbindungen," p. 371. Verlag Chemie, Weinheim, 1966.
40. W. Closson, in Frisch and Wasserman.[3], p. 3791.
40a. V. Prelog and H. Gerlach, *Helv. Chim. Acta* **47**, 2288 (1964); H. Gerlach and J. A. Owtschinnikow, *ibid.* p. 2294.
41. R. Cruse, *in* "Stereochemie der Kohlenstoffverbindungen" (E. L. Eliel, ed.), p. 224. Verlag Chemie, Weinheim, 1966.
42. P. G. Tait, "Collected Works," Vol. 1. Cambridge Univ. Press, London and New York, 1898.
43. K. Reidemeister, *in* "Ergebnisse der Mathematik und ihrer Grenzgebiete," p. 1. Springer, Berlin, 1932.
44. W. J. Ambs, *Mendel Bull.* **17**, 26 (1953).
45. R. L. Kornegay, H. L. Frisch, and E. Wasserman, *J. Org. Chem.* **34**, 2030 (1969).
46. J. G. Kirkwood, *J. Chem. Phys.* **5**, 479 (1937).
47. R. S. Cahn and C. K. Ingold, *J. chem. Soc., London* p. 612 (1951).
48. R. S. Cahn, C. K. Ingold, and V. Prelog, *Experientia* **12**, 81 (1956).
49. R. S. Cahn, C. K. Ingold, and V. Prelog, *Angew. Chem.* **78**, 413 (1966); *Angew. Chem. Int. Ed. Engl.* **5**, 385 (1966).
50. For literature, see W. Bauer and J. Vinograd, *J. Mol. Biol.* **33**, 141 (1968).
50a. R. Radloff, W. Bauer, and J. Vinograd, *Proc. Nat. Acad. Sci. U.S.* **57**, 1514 (1967).
50b. L. Pikó, D. G. Blair, A. Tyler, and J. Vinograd, *Proc. Nat. Acad. Sci. U.S.* **59**, 838 (1968).
51. G. Riou and E. Delain, *Proc. Nat. Acad. Sci. U.S.* **62**, 210 (1969).
52. R. Dulbecco and M. Vogt, *Proc. Nat. Acad. Sci. U.S.* **50**, 236 (1963).
53. R. Weil and J. Vinograd, *Proc. Nat. Acad. Sci. U.S.* **50**, 730 (1963).
54. L. V. Crawford and P. H. Black, *Virology* **24**, 388 (1964).
55. J. Vinograd, J. Lebowitz, R. Radloff, R. Watson, and P. Laipis, *Proc. Nat. Acad. Sci. U.S.* **53**, 1104 (1965).
56. J. Vinograd, J. Lebowitz, and R. Watson, *J. Mol. Biol.* **33**, 173 (1968).

57. Remark of the author, University of Freiburg, 1969.
58. J. Vinograd and J. Lebowitz, *J. Gen. Physiol.* **49**, 103 (1966).
59. L. S. Lerman, *J. Mol. Biol.* **3**, 18 (1961).
60. D. M. Neville and D. R. Davies, *J. Mol. Biol.* **17**, 57 (1966).
61. J.-B. LePecq, Thesis, Faculty of Science, Paris, 1965; J.-B. LePecq and C. Paoletti, *J. Mol. Biol.* **27**, 87 (1967).
62. M. J. Waring, *J. Mol. Biol.* **13**, 269 (1965).
63. W. Fuller and M. J. Waring, *Ber. Bunsenges. Physik. Chem.* **68**, 805 (1964).
64. B. Hudson and J. Vinograd, *Nature (London)* **221**, 332 (1969).
65. R. G. Kostyanovskii, *Khim. Zhizn (Moscow)* p. 34 (1965).
66. C. A. Thomas, *Progr. Nucl. Acid. Res.* **5**, 315 (1966).
67. E. A. Mason, *J. Chem. Phys.* **23**, 49 (1955).
68. K. Ziegler, *in* "Methoden der Organischen Chemie" (E. Müller, ed.), Vol. IV, Part 2, p. 745. Thieme, Stuttgart, 1955.
69. G. Schill, *Chem. Ber.* **98**, 2906 (1965).
70. G. Schill, Dissertation, University of Freiburg, 1959.
71. M. Gardner, *Sci. Amer.* **219**, No. 6, 112 (1968).
72. In an unpublished manuscript, N. van Gulick has dealt with the possibilities of Möbius strips and braids in chemical systems (see footnote 15 in Frisch and Wasserman[3]). We thank the author for making available the manuscript.
73. A. F. Möbius, "Werke," Vol. II. 1858.
74. E. Wasserman, personal communication (1969).
75. A. D. Hershey, E. Burgi, and L. Ingraham, *Proc. Nat. Acad. Sci. U.S.* **49**, 748 (1963).
76. A. D. Hershey and E. Burgi, *Proc. Nat. Acad. Sci. U.S.* **53**, 325 (1965).
77. H. B. Strack and A. D. Kaiser, *J. Mol. Biol.* **12**, 36 (1965).
78. R. L. Baldwin, P. Barrand, A. Fritsch, D. A. Goldthwait, and F. Jacob, *J. Mol. Biol.* **17**, 343 (1966).
79. J. J. Weidmann, H. Kuhn, and W. Kuhn, *J. Chim. Phys.* **50**, 226 (1953).
80. A. Lüttringhaus and H. Preugschas, unpublished data (1959); H. Preugschas, Dissertation, University of Freiburg, 1959.
81. A. Lüttringhaus and G. Schill, unpublished data (1959).
82. A. Lüttringhaus and R. Vollrath, unpublished data (1960).
83. R. Vollrath, Dissertation, University of Freiburg, 1960.
84. A. Lüttringhaus and R. Vollrath, unpublished work (1960–1963).
85. A. Lüttringhaus and G. Isele, unpublished data (1965); G. Isele, Diplomarbeit, University of Freiburg, 1965.
86. H. Zahn and H. Determann, *Chem. Ber.* **90**, 2176 (1957).
87. F. Cramer, "Einschlussverbindungen." Springer, Berlin, 1954.
88. F. Cramer and F. M. Henglein, *Chem. Ber.* **90**, 2561 (1957).
89. K. Ziegler and P. Orth, *Chem. Ber.* **66**, 1867 (1933).
90. H. Stetter and J. Marx, *Justus Liebigs Ann. Chem.* **607**, 59 (1957).
91. G. Schill, *Justus Liebigs Ann. Chem.* **695**, 65 (1966).
92. L. Ruzicka, P. A. Plattner, and W. Widmer, *Helv. Chim. Acta* **25**, 604 (1942).
93. E. J. Salmi, *Ber. Deut. Chem. Ges.* **71**, 1803 (1938).
94. M. Brini, *Bull. Soc. Chim. Fr.* p. 339 (1955).
95. V. Prelog, M. M. Wirth, and L. Ruzicka, *Helv. Chim. Acta* **29**, 1425 (1946).
96. See K. S. Pitzer, *Science* **101**, 672 (1945); G. Bier, *Experientia* **2**, 82 (1946).
97. D. P. Evans, *J. Chem. Soc., London* p. 785 (1936); D. P. Evans and J. J. Gordon, *ibid.* p. 1434 (1938).

98. V. Prelog, L. Ruzicka, and O. Metzler, *Helv. Chim. Acta* **30**, 1883 (1947).

99. V. Prelog, L. Ruzicka, P. Barman, and L. Frenkiel, *Helv. Chim. Acta* **31**, 92 (1948).

99a. V. Prelog, K. Wiesner, W. Ingold, and O. Häfliger, *Helv. Chim. Acta* **31**, 1325 (1948).

100. V. Prelog, P. Barman, and M. Zimmermann, *Helv. Chim. Acta* **32**, 1284 (1949).

101. V. Prelog, P. Barman, and M. Zimmermann, *Helv. Chim. Acta* **33**, 356 (1950).

101a. T. Ledaal, *Tetrahedron Lett.* p. 651 (1968).

102. N. L. Allinger and J. J. Maul, *Tetrahedron* **24**, 4257 (1968).

103. M. Stoll, J. Hulstkamp, and A. Rouvé, *Helv. Chim. Acta* **31**, 543 (1948).

104. V. Prelog and K. Wiesner, *Helv. Chim. Acta* **30**, 1465 (1947).

105. L. Ruzicka and G. Giacomello, *Helv. Chim. Acta* **20**, 548 (1937).

105a. G. Salomon, *Helv. Chim. Acta* **19**, 743 (1936); *Trans. Faraday Soc.* **32**, 153 (1936).

105b. A. Lüttringhaus and R. Kohlhaas, *Ber. Deut. Chem. Ges.* **72**, 907 (1939); A. Lüttringhaus, *Naturwissenschaften* **30**, 40 (1942); A. Lüttringhaus and I. Sichert-Modrow, *J. Makromol. Chem.* **18/19**, 511 (1956).

106. G. Wittig, *in* "Neuere Methoden der präparativen organischen Chemie," (W. Foerst, ed.), Vol. 1, p. 476. Verlag Chemie, Weinheim, 1943.

107. G. Schill, *Chem. Ber.* **99**, 2689 (1966).

108. Huang-Minlon, *J. Amer. Chem. Soc.* **68**, 2487 (1946).

109. J. B. Menke, *Rec. Trav. Chim. Pays-Bas* **44**, 141 and 269 (1925).

110. K. Ziegler, K. Schneider, and J. Schneider, *Justus Liebigs Ann. Chem.* **623**, 9 (1959).

111. A. Lüttringhaus and S. Linke, unpublished data (1964); S. Linke, Dissertation, University of Freiburg, 1964.

112. K. Ziegler and H. Weber, *Ber. Deut. Chem. Ges.* **70**, 1275 (1937).

113. A. Lüttringhaus and H. Simon, *Justus Liebigs Ann. Chem.* **557**, 120 (1947).

114. W. S. Fones, *J. Org. Chem.* **14**, 1099 (1949); R. C. Fuson and H. O. House, *J. Amer. Chem. Soc.* **75**, 1327 (1953).

115. W. E. Truce and F. E. Roberts, *J. Org. Chem.* **28**, 961 (1963).

116. A. Lüttringhaus and K. Steigerwald, unpublished data (1964); K. Steigerwald, Dissertation, University of Freiburg, 1965.

117. C. R. Hauser, F. W. Swamer, and B. I. Ringler, *J. Amer. Chem. Soc.* **70**, 4023 (1948).

118. S. Hünig, E. Lücke, and E. Benzing, *Chem. Ber.* **91**, 129 (1958); S. Hünig and W. Eckardt, *ibid.* **95**, 2493 (1962).

119. H. Kessler and A. Rieker, *Justus Liebigs Ann. Chem.* **708**, 57 (1967).

120. See A. Lüttringhaus.[17] p. 73, footnote 38b.

121. A. Lüttringhaus and J. Winterhalter, unpublished data (1960); J. Winterhalter, Dissertation, University of Freiburg, 1960.

121a. J. C. Sauer, *J. Amer. Chem. Soc.* **69**, 2444 (1947).

122. E. Müller and W. Rundel, *Angew. Chem.* **70**, 105 (1958).

123. A. Lüttringhaus and S. Linke, unpublished work (1964–1965).

124. R. Schröter, *in* "Methoden der organischen Chemie" (E. Müller, ed.), Vol. XI, Part 1, p. 788. Thieme, Stuttgart, 1957.

125. L. Friedman and H. Shechter, *J. Org. Chem.* **26**, 2522 (1961).

126. G. Schill and L. Tafelmair, to be published; L. Tafelmair, Dissertation, University of Freiburg, 1969.

127. H. C. Brown, K. J. Murray, L. J. Murray, J. A. Snover, and G. Zweifel, *J. Amer. Chem. Soc.* **82**, 4233 (1960).

128. H. C. Brown and P. Heim, *J. Amer. Chem. Soc.* **86**, 3566 (1964).

129. J. von Braun, K. Heider, and E. Müller, *Ber. Deut. Chem. Ges.* **51**, 737 (1918).

130. R. Munch, G. T. Thannhauser, and D. L. Cottle, *J. Amer. Chem. Soc.* **68**, 1297 (1946).
131. T. Doornbos and J. Strating, *Rec. Trav. Chim. Pays-Bas* **85**, 41 (1966).
132. D. I. Weisblat, B. J. Magerlein, and D. R. Myers, *J. Amer. Chem. Soc.* **75**, 3630 (1953).
133. H. Stetter and E. E. Roos, *Chem. Ber.* **87**, 566 (1954).
134. H. R. Snyder and R. E. Heckert, *J. Amer. Chem. Soc.* **74**, 2006 (1952); H. R. Snyder and H. C. Geller, p. 4864.
135. H. Stetter, *Chem. Ber.* **85**, 197 and 380 (1953).
136. R. C. Fuson and H. O. House, *J. Amer. Chem. Soc.* **75**, 1327 and 5744 (1953).
137. K. Wiesner and D. E. Orr, *Tetrahedron Lett.* No. 16, p. 11 (1960).
138. H. Stetter and E. E. Roos, *Chem. Ber.* **88**, 1390 (1955).
139. H. Stetter and K. H. Mayer, *Chem. Ber.* **94**, 1410 (1961).
140. H. Stetter and L. Marx-Moll, *Chem. Ber.* **91**, 677 (1958).
141. H. Stetter, L. Marx-Moll, and H. Rutzen, *Chem. Ber.* **91**, 1775 (1958).
142. M. Rothe and R. Timler, *Chem. Ber.* **95**, 783 (1962).
143. J. Dale and R. Coulon, *J. Chem. Soc.* p. 182 (1964).
144. Z. B. Papanastassiou and R. J. Bruni, *J. Org. Chem.* **29**, 2870 (1964).
145. S. Corsano and F. Bombardiere, *Ann. Chim.* (*Rome*) **54**, 650 (1964); *Chem. Abstr.* **61**, 13141 (1964).
146. G. Schill, *Chem. Ber.* **98**, 3439 (1965).
147. S. Hünig and P. Richters, *Chem. Ber.* **91**, 442 (1958).
148. S. Dähne, *Z. Chem.* **3**, 191 (1963).
149. A. J. Birch, *J. Chem. Soc.* (*London*) p. 593 (1946); p. 102 (1947).
150. See A. J. Birch and H. Smith, *Quart. Rev.* (*London*), **12**, 32 (1958).
151. B. Emmert, *Ber. Deut. Chem. Ges.* **42**, 1507 (1909).
152. L. Horner and A. Mentrup, *Justus Liebigs Ann. Chem.* **646**, 49 (1961).
153. R. Kuhn and H. J. Haas, *Justus Liebigs Ann. Chem.* **611**, 57 (1958).
154. P. D. Swaters, unpublished data (see Doornbos Ph. D. Thesis[38], p. 85).
155. F. Kehrmann, *Ber. Deut. Chem. Ges.* **23**, 905 (1890).
156. W. K. Anslow and H. Raistrick, *J. Chem. Soc.* p. 1446 (1939).
157. G. Schill and H. Neubauer, to be published.
158. S. Petersen, W. Gauss, and E. Urbschat, *Angew. Chem.* **67**, 217 (1955).
159. G. Schill and H. Neubauer, unpublished data (1965).
160. G. Schill and H. Neubauer, *Synthesis* **1**, 77 (1969).
161. G. Schill, *Justus Liebigs Ann. Chem.* **691**, 79 (1966).
162. G. Schill, *Justus Liebigs Ann. Chem.* **693**, 182 (1966).
163. J. Kollonitsch, O. Fuchs, and V. Gábor, *Nature* (*London*) **175**, 346 (1955).
164. D. Wasserman and C. R. Dawson, *J. Amer. Chem. Soc.* **72**, 4994 (1950).
165. G. Schill, *Justus Liebigs Ann. Chem.* **695**, 65 (1966).
166. G. Schill and R. Henschel, *Justus Liebigs Ann. Chem.* **731**, 113 (1970).
167. G. Slooff, *Rec. Trav. Chim. Pays-Bas* **54**, 995 (1935).
168. L. F. Fieser and M. I. Ardao, *J. Amer. Chem. Soc.* **78**, 774 (1956).
169. L. Horner, L. Schläfer, and H. Kämmerer, *Chem. Ber.* **92**, 1700 (1959).
170. E. Hoehn, *Helv. Chim. Acta*, **8**, 275 (1925).
171. W. K. Anslow and H. Raistrick, *J. Chem. Soc.* (*London*) p. 1446 (1939).
172. W. Brackman and E. Havinga, *Rec. Trav. Chim. Pays-Bas* **74**, 937 (1955).
173. L. Horner and K. Sturm, *Chem. Ber.* **88**, 329 (1955).

174. W. Draber, Dissertation, University of Freiburg, 1960.
175. E. Adler and B. Stenemur, *Chem. Ber.* **89**, 291 (1956).
176. P. da Re and L. Cimatoribus, *J. Org. Chem.* **26**, 3650 (1961).
177. L. D. Bergelson and M. M. Schemjakin, *Angew. Chem.* **76**, 113 (1964); *Angew. Chem. Int. Ed. Engl.* **3**, 250 (1964).
178. N. Petragnani and G. Schill, *Chem. Ber.* **97**, 3293 (1964).
179. H. Meerwein, *Ber. Deut. Chem. Ges.* **66**, 411 (1933).
180. C. D. Hurt and W. A. Hoffman, *J. Org. Chem.* **5**, 212 (1940).
181. J. S. Byck and C. R. Dawson, *J. Org. Chem.* **33**, 2451 (1968).
182. J. J. Pappas, W. P. Keaveney, E. Gancher, and M. Berger, *Tetrahedron Lett.* 4273 (1966).
183. U. Schräpler and R. Rühlmann, *Chem. Ber.* **97**, 1383 (1964).
184. J. S. Byck and C. R. Dawson, *J. Org. Chem.* **42**, 1084 (1967).
185. G. Schill, W. Vetter, and K. Murjahn, *Justus Liebigs Ann. Chem.*, in press; K. Murjahn, Dissertation, University of Freiburg, 1969.
186. H. Gilman, J. Swiss, and L. C. Cheney, *J. Amer. Chem. Soc.* **62**, 1963 (1940).
187. D. Papa, E. Schwenk, and H. Hankin, *J. Amer. Chem. Soc.* **69**, 3018 (1947).
188. G. Schill and C. Zürcher, to be published; part of the not yet finished dissertation of C. Zürcher.
189. G. Schill and C. Zürcher, to be published; part of the Diplomarbeit, C. Zürcher, University of Freiburg, 1968.
190. H. Gross and J. Gloede, *Chem. Ber.* **96**, 1387 (1963).
191. H. Gross and U. Karsch, *J. Prakt. Chem.* [4] **29**, 315 (1965).
192. I. M. Downie, J. B. Holmes, and J. B. Lee, *Chem. Ind. (London)* p. 900 (1966); J. B. Lee and I. M. Downie, *Tetrahedron* **23**, 359 (1967).
193. L. Horner, H. Oediger, and H. Hoffmann, *Justus Liebigs Ann. Chem.* **626**, 26 (1959).
194. J. F. W. McOmie and M. L. Watts, *Chem. Ind. (London)* p. 1658 (1963).
195. W. Flaig and J. C. Salfeld, *Justus Liebigs Ann. Chem.* **618**, 117 (1958).
196. H. Hoyer, *Angew. Chem.* **73**, 465 (1961).
197. G. Schill and C. Zürcher, *Angew. Chem.* **81**, 996 (1969); *Angew. Chem. Int. Engl. Ed.* **8**, 988 (1969).
198. R. B. Merrifield, *J. Amer. Chem. Soc.* **85**, 2149 (1963).
199. H. Stetter and K. Swincicki, unpublished data (1957); K. Swincicki, Dissertation, University of München, 1957.

Author Index

Numbers in parentheses are reference numbers and indicate that an author's work is referred to, although his name is not cited in the text. Numbers in italics show the page on which the complete reference is listed.

A

Adler, E., 103(175), 104(175), 108(175), *181*
Allinger, N. L., 42(102), *179*
Ambs, W. J., 15(44), *177*
Anslow, W. K., 91(156), 102(171), *180*
Ardao, M. I., 100(168), *180*
Aurnhammer, R., 2(8), 44(8), 105(8), *176*

B

Baldwin, R. L., 36(78), *178*
Barman, P., 2(12), 34(12), 37(12), 42(99, 100, 101), 49(12), 105(12), *176, 179*
Barrand, P., 36(78), *178*
Bauer, W., 19(50, 50a), 20(50a), *177*
Benzing, E., 57(118), *179*
Bergelson, L. D., 104(177), 105(177), *181*
Berger, M., 105(182), *181*
Bier, G., 41(96), *178*
Birch, A. J., 89(149, 150), *180*
Black, P. H., 19(54), *177*
Blair, D. G., 19(50b), 21(50b), 175(50b), *177*
Boeckmann, J., 4(30, 33), 157(30), 164(30), 170(30), *177*
Bombardiere, F., 82(145), 95(145), *180*
Brackman, W., 102(172), *180*
Brini, M., 41(94), *178*
Brown, C. J., 1(4), 34(4), *176*
Brown, H. C., 66(127, 128), 82(128), 95(128), *179*
Bruni, R. J., 82(144), *180*
Burgi, E., 36(75, 76), *178*
Byck, J. S., 105(181), 107(184), *181*

C

Cahn, R. S., 17(47, 48, 49), *177*
Cheney, L. C., 107(186), *181*
Cimatoribus, L., 103(176), *181*
Clayton, D. A., 3(24), 20(24), 21(24), 175(24), *176*
Closson, W., 12(40), *177*
Coburn, E. R., 1(4), 34(4), *176*
Corsano, S., 82(145), 95(145), *180*
Cottle, D. L., 73(130), *180*
Coulon, R., 82(143), *180*
Cramer, F., 1(5), 2(5), 3(27), 26(5), 38(5, 87, 88), 145(27), 146(27), *176, 177, 178*
Crawford, L. V., 19(54), *177*
Cruse, R., 12(41), *177*

D

Dähne, S., 89(148), *180*
Dale, J., 82(143), *180*
da Re, P., 103(176), *181*
Davies, D. R., 20(60), *178*
Dawson, C. R., 99(164), 105(181), 107(184), *180, 181*
Delain, E., 19(51), 21(51), 175(51), *177*
Derst, P., 2(7), *176*
Determann, H., 38(86), *178*
Dieterich, D., 7(37), *177*
Doornbos, T., 11(38), 12(38), 13(38), 78(131), 79(38), 80(38), 82(38), 84(38), 86(38, 131), 87(38), 88(38), 89(38), 90(38), 93(38), 110(38), 167(38, 131), 168(38, 131), 171(38), *177, 180*

183

Subject Index

A

Acetolysis of *N*,*N*-dialkylaminomethyl-hydroquinone ethers, 66, 67, 72, 150

Acid halide–amine cyclization to macrocyclic diamides, 40, 82

Acyloin cyclization
  to ansa compounds, 70, 99
  failure of, 49, 105
  of ketals, 44
  to macrocycles, 35, 73

Acyloin substituents, Clemmensen reduction of, 70, 73

Alkylation
  of acylanilides, 53, 57
  of aniline derivatives, 53, 57, 73, 79–81, 83, 87, 118, 154
  of 1,4-*N*,*N*-dialkylamino-2,5-dihydroxy-benzenes, 86

Allyl rearrangement of guaiacol ethers, 104, 105

Ansa compounds
  of benzene, 146
  of catechol, 106
  of 1,4-diaminobenzene, 82
  of 1,4-diamino-2,5-dibrombenzene, 81, 82
  of 1,4-diamino-2,5-dimethoxybenzene, 81, 82
  of hydroquinone, 47, 99
  of resorcinol, 47
  of veratrole, 51, 105, 106, 111

Azacycloketones, 102, 117, 120

B

Benzodioxoles, 58, 60, 61, 99–102, 108–110, 112–120, 132–137, 153, 154, 164, 165

Benzoquinones
  2,5-dialkyl-, 95
  2,5-*N*-*N*-dialkylamino-3,6-dialkyl-, 97

Benzoquinones—*cont.*
  2,5-dialkyl-3,6-dinitro-, 95, 97
  2,5-dihydroxy-3,6-dimethyl-, 97, 119, 120
  2-hydroxy-3,5-dialkyl-, 102, 103
  2-hydroxy-3,5-polymethylene-, 57, 118
  2-methoxy-5-hydroxy-3,6-dimethyl-, 119, 120

Birch reduction, failure with diansa compounds, 89

Bisdiansa compounds, 131–137

Borromean rings, stereochemistry of, 14

C

Catechols
  3,5-dialkyl-, preparation of, 103–107
  3,5-polymethylene-, 106

Catenanes
  absolute configuration of, 17, 18
  higher linear
    concept of synthesis, 162
    model investigations, 164
  history of synthetic work on, 1, 2
  mass spectrum of, 121, 122
  naturally occurring, 3, 19–21
  nomenclature of, 7–10
  in polymers, speculations, 1, 2
  stereochemistry of, 11–14

[2]-Catenanes
  arguments for structure of, 116, 117, 121
  calculations on formation probability, 37
  cycloenantiomerism in, 12
  directed syntheses, 29, 114
  doubly wound, concept of synthesis, 159
  mass spectrum of, 121
  methods for syntheses, 25
  ring size requirements, 25
  statistical syntheses, 34

189